Classical Mechanics

Taeseong Jeong

Contents

Imagination is more important than knowledge

– Albert Einstein

Chapter 1

Introduction

1.1 Degree of Freedom

The number of independent components of the configuration q is called the number of degrees of freedom of the mechanical system. The number of degrees of freedom is the number of independent ways in which the system can move. Here are some examples. A point mass sliding on an airtrack, described by only a single coordinate, has one degree of freedom. We get more degree of freedom if we go to more dimensions or to more complicates objects.

1.2 Space and Time

The most fundamental assumptions of physics are the concepts of space and time. We assume that space and time are continuous, that it is meaningful to say that an event occurred at a specific point in space and specific time. These assumptions are common to the whole of physics and there is no convincing evidence that we have as yet reached the limits of their range of validity.

In classical physics, we assume further that there is a universal time scale, that the geometry of space is Euclidean, and that there is no limit in principle to the accuracy with which we can measure positions and velocities. These assumptions have been somewhat modified in quantum mechanics and relativity. However we shall take them for granted here and concentrate our attention on the more specific assumptions of classical mechanics.

1.3   Inertial frame

It is useful to introduce the concept of a frame of reference. To specify positions and times, each observer may choose a zero of the time scale, an origin in space, and a set of three Cartesian coordinate axes. We shall refer to these collectively as a frame of reference. The position and time of any event may then be specified with respect to this frame by the three Cartesian coordinates, x, y, z and the time t. We may suppose that the observer is located on a solid body, such as the earth, that he chooses some point of this body as his origin, and takes his axes to be rigidly fixed to it.

In view of the relativity principle, the frames of reference used by different un-accelerated observers are completely equivalent. The laws of physics expressed in terms of x, y, z, t must be identical with those in terms of the coordinates of another frame, x', y', z', t'. They are not identical with the laws expressed in terms of the coordinates used by an accelerated observer. The frames used by unaccelerated

observers are called "inertial frames".

We need a criterion to distinguish inertial frames from the others. Formally, an

inertial frame may be defined to be one with respect to which an isolated body, far

removed from all other matter, would move with uniform velocity. This is of course

an idealized definition, since in practice we never can get infinitely far away from

other matter. For all practical purposes an inertial frame is one whose orientation is

fixed relative to the 'fixed stars', and in which the sun moves with uniform velocity.

It is an essential assumption of classical mechanics that such frames exist. Indeed

this assumption is the real physical content of Newton's first law.

1.4   Vectors

It is very convenient to use a notation which does not refer explicitly to a particular

set of coordinate axes. Instead of using Cartesian coordinates x, y, z, we may specify

the position of a point P with respect to a given origin O by the length and direction

of the line OP. A quantity which is specified by a magnitude and a direction is

called a vector; in this case the position vector $\mathbf{r}$ of P with respect to O. Many other

physical quantities are also vectors: examples are velocity and force. They are to be

distinguished from scalars (like mass and energy) which are completely specified by

a magnitude alone.

The magnitude of the vector will be denoted by the corresponding $|\mathbf{r}|$. The scalar

and vector products of two vectors $\mathbf{a}$ and $\mathbf{b}$ will be written $\mathbf{a} \cdot \mathbf{b}$ and $\mathbf{a} \times \mathbf{b}$ respectively.

We shall use $\hat{r}$ to denote the unit vector in the direction of $\mathbf{r}$, $\hat{r} = \frac{\mathbf{r}}{r}$. The unit vectors along the x, y, z axes will be denoted by $\mathbf{i}, \mathbf{j}, \mathbf{k}$, so that

$$\mathbf{r} = x\mathbf{i} + y\mathbf{j} + z\mathbf{k} \tag{1.1}$$

We shall use the vector notation in formulating the basic laws of mechanics, both because of the mathematical simplicity and because of the physical ideas behind the mathematical formalism are often much clearer in terms of vectors.

1.5 The relativity principle

In Newtonian mechanics bodies fall downwards because they are attracted towards the earth, rather than towards some fixed point in space. Thus position has a meaning only relative to the earth, or to some other body. In the same way, velocity has only a relative significance. Given two bodies movig with uniform relative velocity, it is impossible in principle to decide which of them is at rest, and which is moving. This statement is the principle of relativity.

However, acceleration still remains an absolute meaning, since it is experimentally possible to distinguish between motion with uniform velocity and accelerated motion. If we are sitting inside an aircraft, we can easily detect its acceleration, but we cannot measure its velocity.

If two unaccelerated observers perform the same experiment, they must arrive at the same results. It makes no difference whether it is performed on the ground or in a smoothly travelling vehicle. However, if an accelerated observer performs the experiment, he may well get a different answer. The relativity principle asserts that all unaccelerated observers are equivalent.

1.6   Motion, trajectories, and Newton's Laws

A trajectory is a possible motion of a physical system. We describe a trajectory by giving the value, $q(t)$ of the coordinate of the system as a function of time. There are an infinite number of possible trajectories for any interesting physical system. Our everyday experience tells us that knowing the configuration of the system at some time is not enough information to allow us to calculate the trajectory. A snapshot does not tell us how the system is moving. We also need to know the velocity of the system, the derivative of q with respect to time.

$$\dot{q}(t) \equiv \frac{d}{dt} q(t). \tag{1.2}$$

If q describe more than one coordinate, then $\dot{q}$ describes the same number of independent velocities-but we will continue to call $\dot{q}$ the velocity, even if it has multiple components. If q is a vector, $\dot{q}$ is the same kind of vector. For the particle

$$q = (x, y), \dot{q} = (\dot{x}, \dot{y}). \tag{1.3}$$

Then we can say that if we know q and $\dot{q}$ at a particular time, we should be able to calculate the trajectory.

One of the central problems of mechanics is to calculate what these trajectories are for various systems. To begin let us think about what this process is like for a system specified by a single coordinate, x. Newton tells us that to figure out how x changes as a function of time, we define our coordinates in an appropriate inertial frame and use his second law

$$\mathbf{F} = m\mathbf{a} \tag{1.4}$$

where F is the force on the objects, m is its mass, and a is its acceleration.

$$a \equiv \frac{d^2}{dt^2}x \equiv \ddot{x}. \tag{1.5}$$

For now we will just assume that m is a fixed property of the object. In general force F will depend on what the system is doing. Since we have already assumed that all we need to know about the system at a given time is q and $\dot{q}$, in this case x and $\dot{x}$, all F can depend on is $q, \dot{q}$, and t.

So in general, F at time t is some function of x and $\dot{x}$, at that time, and $t, F(x(t), \dot{x}(t), t)$. Then Newton's second law becomes a formular for the acceleration:

$$\ddot{x} \equiv \frac{1}{m}F(x, \dot{x}, t). \tag{1.6}$$

Newton's second law for a single particle of mass m can be written as

$$\mathbf{F} = \frac{d\mathbf{p}}{dt}. \tag{1.7}$$

where the quantity $\vec{p}$ is the momentum of the particle, and is given in Newtonian mechanics by

$$\mathbf{p} = m\mathbf{v}. \tag{1.8}$$

The equation of motion, which specifies how the body will move is Newton's second law (mass× acceleration =force).

$$\dot{\mathbf{p}}_i = m_i \mathbf{a}_i = \mathbf{F}_i \tag{1.9}$$

where $\mathbf{F}_i$ is the total force acting on the body. This force is composed of a sum of forces due to each of the other bodies in the system. If we denote the force on the ith body due to the jth body by $\mathbf{F}_{ij}$, then

$$\mathbf{F}_i = \mathbf{F}_{i1} + \mathbf{F}_{i2} + \cdots + \mathbf{F}_{iN} = \sum_{j=1}^{N} \mathbf{F}_{ij} \tag{1.10}$$

where, of course, $\mathbf{F}_{ii} = 0$, since there is no force on the ith body due to itself. Note that since the right side of 1.10 is a vector sum, this equation incorporates the "parallelogram law" of composition of forces.

The two-body forces $\mathbf{F}_{ij}$ must satisfy Newton's third law, which asserts that "action" and "reaction" are equal and opposite,

$$\mathbf{F}_{ji} = -\mathbf{F}_{ij} \tag{1.11}$$

Moreover, $\mathbf{F}_{ij}$ is a function of the positions and velocities oft he ith and jth bodies, but is unaffected by the presence of other bodies.

So if force is always changing momentum according to (1), how is it that momentum is conserve? The answer that you probably learned in high-school is Newton's third law. For every action, there is an equal and opposite reaction. If thing 1 produces a force on thing 2, then thing 2 produces a force with equal magnitude and in the opposite direction on thing 1. If this is correct, then any change of the momentum of something is always compensated by a change in the momentum of the things that are producing the forces on it. The total momentum of any isolated system that has no external forces acting on it is always conserved.

According to Newton's law of universal gravitation, there is a force of this type between every pair of bodies, proportional in magnitude to the product of their masses. It is given by

$$f(r_{ij}) = -\frac{Gm_i m_j}{r_{ij}^2},$$ (1.12)

where G is the gravitational constant, whose value is

$$G = 6.67 \times 10^{-11} Nm^2 kg^{-2}$$ (1.13)

Since the masses are always positive, this force is always attractive.

In addition, if the bodies are electrically charged, there is an electrostatic force given by

$$f(r_{ij}) = \frac{q_i q_j}{4\pi\epsilon_0 r_{ij}^2}$$ (1.14)

where q_i and q_j are their electric charges and ϵ_0 is another constant,

$$\epsilon_0 = 8.854 \times 10^{-12} Fm^{-1}$$ (1.15)

Note that the analogue of Newton's constant G is

$$\frac{1}{4\pi\epsilon_0} = 8.99 \times 10^9 Nm^2 C^{-2}$$ (1.16)

Electric charges may be of either sign, and therefore the electrostatic force may be repulsive or attractive according to the relative sign of q_i and q_j. Note that the enormous difference in the orders of magnitude of the constants G and $\frac{1}{4\pi\epsilon_0}$ when expressed in SI units. This means that the gravitational forces are really exceptionally weak. They appear significant to us only because we happen to leave close to a body of very large mass. Corresponding large charges do not appear, because positive and negative charges largely cancel out, leaving macroscopic bodies with a net charge close to zero.

1.7   Mass and force

It is an important principle of physics that no quantity should be introduced into the theory which cannot be measured. Newton's laws involve not only the concepts of velocity and acceleration, which can be measured by measuring distances and times, but also the new concepts of mass and force. To give the laws a physical meaning we have to show that these are measurable quantities. This is not quite as trivial as it might seem at first sight, for it is easy to see that any experiment designed to measure these quantities must necessarily involve Newton's laws themselves in its interpretation.

Let us consider the measurement of mass. Since the units of mass are arbitrary, we have to specify a way of comparing the masses of two given bodies. It is important to realize that we are discussing here the inertial mass, which appears in Newton's second law. and not the gravitational mass. The two are proportional, but this is a physical law derived from experimental observation rather than an a priori assumption.

Clearly we can compare the inertial masses of two bodies by subjecting them to equal forces and comparing their accelerations, but this does not help unless we have some way of knowledge that the forces are equal. However, there is one case in which we do know this, because of Newton's third law. If we isolate the two bodies from all other matter, and compare their mutually induced accelerations, then

$$m_1 \mathbf{a}_1 = -m_2 \mathbf{a}_2 \tag{1.17}$$

so that the accelerations are oppositely directed, and inversely proportional to the masses. This is one of the commonly used methods for comparing masses. If we allow two small bodies to collide, then during the collision the effects of more remote bodies are generally negligible in comparison with their effect on each other, and we may treat them approximately as an isolated system. The mass ratio can then be determined from measurements of their velocities before and after the collision, then the law of conservation of momentum,

$$m_1 \mathbf{v}_1 + m_2 \mathbf{v}_2 = constant \tag{1.18}$$

If we wish to separate the definition of mass from the physical content of equation 1.17, we may adopt as a fundamental law the following

In an isolated two-body system, the accelerations always satisfy the relation $\mathbf{a}_1 = -k_{21}\mathbf{a}_2$, *where the scalar* k_{21} *is, for two given bodies, a constant independent of their positions, velocities and inertial states.*

If we choose the first body to be a standard body, and conventionally assign it unit mass, then we may define the mass of the second body to be $m_2 = k_{21}$ in units

of this standard mass.

Chapter 2

Linear motion

Let's discuss the motion of a body which is free to move only in one dimension. The problems considered are chosen to illustrate the concepts and techniques which will be of use in the more general case of three dimensional motion.

2.1 Conservative forces

We consider a particle moving long a line, under a force which is given as a function of its position, $F(x)$. Then the equation of motion is

$$m\ddot{x} = F(x) \tag{2.1}$$

Since this equation is of second order in the time derivatives, we shall have to integrate twice to find x as a function of t. Thus the solution will contain two arbitrary constants that may be fixed by specifying the initial values of x and $\dot{x}$.

When the force depends only on x, we can always find a first integral, a function of x and $\dot{x}$ whose value is constant in time. Let us consider the kinetic energy.

$$T = \frac{1}{2}m\dot{x}^2 \tag{2.2}$$

Differentiating, we find for its rate of change

$$\dot{T} = m\dot{x}\ddot{x} = F(x)\dot{x} \tag{2.3}$$

Integrating with respect to time, we find

$$T = \int F(x)\dot{x}dt = \int F(x)dx, \tag{2.4}$$

If we define the potential energy

$$V(x) = -\int_{x_0}^{x} F(x)dx, \tag{2.5}$$

We can write 2.4 in the form

$$T + V = E = const \tag{2.6}$$

Here x_0 is an arbitrary constant, corresponding to the arbitrary additive constant of integration. There is a corresponding arbitrariness up to an additive constant in the total energy E.

Note that equation 2.5 can be in inverted to give the force in terms of the potential energy

$$F(x) = -\frac{dV}{dx} \tag{2.7}$$

The equation 2.6 is the law of conservation of energy. A force of this type, depending only on x, is called a conservative force. A great deal of information about the motion can be obtained form this conservation law, even without integrating again to find x explicitly as a function of t. If the initial position and velocity are given, we can calculate the value of the constant E. Then the equation 2.6 in the form

$$\frac{1}{2}m\dot{x}^2 = E - V \tag{2.8}$$

gives the velocity of the particle when it is at any given position x. Since the kinetic energy is obviously positive, the motion is confined to the region where

$$V(x) \leq E. \tag{2.9}$$

To illustrate these ideas, let us consider the problem of a simple pendulum, comprising a bob of mass m supported by a light rigid rod of length l. The distance x moved by the bob is $x = l\theta$, where θ is the angular displacement. Corresponding tot he restoring force $F = -mg\sin\theta = -mg\sin(\frac{x}{l})$, the potential energy function is

$$V = mgl[1 - \cos(\frac{x}{l})] = mgl[1 - \cos\theta] \tag{2.10}$$

Suppose that initially $\theta = 0$, and that the bob is given a push that starts it moving with velocity v. Then since we have chosen the arbitrary constant in V so that $V(0) = 0$, the total energy is $E = \frac{1}{2}mv^2$. Now if $E < 2mgl$, the motion will be confined between two angles $\pm\theta_0$, where $V(\theta_0) = E$,

$$1 - \cos \theta_0 = \frac{v^2}{2gl} \tag{2.11}$$

These are the points where the kinetic energy vanishes, so that that pendulum bob is instantaneously at rest. The motion is an oscillation of amplitude θ_0.

On the other hand, if the bob is pushed so hard that $E > 2mgl$, then the kinetic energy will never vanish. The bob still has positive kinetic energy when it reaches the upward vertical, namely

$$\frac{1}{2}mv'^2 = E - 2mgl = \frac{1}{2}mv^2 - 2mgl \tag{2.12}$$

In this case the motion is a continuous revolution rather than an oscillation.

2.2   The harmonic Oscillator

A particle can be in equalbrium only if the force acting on it is zero. For a conservative force that the potential energy curve is horizontal at the position of the particle. Let us consider the motion of a particle near a position of equilibrium. Without loss of generality, we may choose the equilibrium point to be the origin $x = 0$, and choose the arbitrary constant in V so that $V(0) = 0$. If we are interested only in small displacements, we may expand $V(x)$ in a Maclaurin-Taylor series,

$$V(x) = V(0) + xV'(0) + \frac{1}{2}x^2V''(0) + \cdots \tag{2.13}$$

where the primes denote derivatives with respect to x. Since we have chosen $V(0) = 0$, the constant term is absent, and since the equalibrium condition is $V'(0) =$

0, the linear term is absent also. Thus near $x = 0$ we can write, approximately,

$$V(x) = \frac{1}{2}kx^2, k = V''(0), \tag{2.14}$$

Because motion near almost any point of equalbrium is described approximately by this potential energy function, it is remarkably ubiquitous. It will therefore be useful to analyze it in some detail. The potential energy curve here is a parabola. Thus for $k > 0$ and any energy $E > 0$, there are two points where $V(x) = E$, namely

$$x = \pm a, a = (2E/k)^{1/2} \tag{2.15}$$

The motion is an oscillation between these points.

On the other hand, if $k < 0$ the curve is an inverted parabola. In this case, two kinds of motion are possible. If $E < 0$, the particle may approach to some minimum distance, where it comes momentarily to rest before reversing direction. But if $E > 0$, it has enough energy to surmount the barrier and will never come to rest.

The force corresponding to the potential energy function is, by

$$F(x) = -kx \tag{2.16}$$

It is an attractive or repulsive force, according as $k.0$ or $k < 0$, proportional to displacement form the point of equalbrium.

The equation of motion may be written

$$m\ddot{x} + kx = 0. \tag{2.17}$$

This equation is very easy to solve. We could proceed directly from the energy conservation equation in the form 2.8 , solving for $\dot{x}$ and integrating again to obtain

$$\int (\frac{2E}{m} - \frac{k}{m}x^2)^{-\frac{1}{2}}dx = \int dt \tag{2.18}$$

However, since we shall encounter a number of similar equations later, it will be useful to discuss an alternative method of solution that can be adapted later to other examples.

Equation 2.17 is a linear differential equation; that is, one involving only linear terms in x and its derivatives. Such equations have the important property that their solutions satisfy the superposition principle; if $x_1(t)$ and $x_2(t)$ are solutions, then so is any linear combination

$$x(t) = a_1 x_1(t) + a_2 x_2(t) \tag{2.19}$$

where a_1 and a_2 are constants

$$m\ddot{x} + kx = a_1(m\ddot{x}_1 + kx_1) + a_2(m\ddot{x}_2 + kx_2) = 0 \tag{2.20}$$

Moreover, if x_1 and x_2 are linearly independent solutions, then 2.19 is actually the general solution. Since equation 2.17 is of second order, we could obtain its solution by integrating twice, and the general solution must therefore contain just two arbitrary constants of integration. All we have to do is to find any two independent solutions $x_1(t)$ and $x_2(t)$.

Let us consider first the case where $k < 0$, so that $V(x)$ has a maximum at $x = 0$. Then 2.17 can be written

$$\ddot{x} - p^2 x = 0, p = (-\frac{k}{m})^{1/2} \tag{2.21}$$

It is easy to verify that this equation is satisfied by the functions $x = e^{pt}$ and $x = e^{-pt}$. Thus the general solution is

$$x = \frac{1}{2}Ae^{pt} + \frac{1}{2}Be^{-pt} \tag{2.22}$$

Clearly, a small displacement will in general lead to an exponential increase of x with time, which continues until the approximation involved in 2.14 ceases to be valid. Thus the equilibrium is unstable, as we should expect when V has a maximum.

We now turn to the case where $k > 0$, and $V(x)$ has a minimum at $x = 0$. Then the potential energy function 2.14 is that of a simple harmonic oscillator. The equation of motion becomes

$$\ddot{x} + \omega x = 0, \omega = (k/m)^{1/2} \tag{2.23}$$

It is again very easy to check that the functions $x = \cos \omega t$ and $x = \sin \omega t$ are solutions of this equation, and the general solution is therefore

$$x = c \cos \omega t + d \sin \omega t \tag{2.24}$$

The arbitrary constants c and d are to be determined by the initial conditions. If at $t = 0$ the particle is at x_0, with velocity v_0, then we easily find

$$c = x_0, d = \frac{v_0}{\omega} \tag{2.25}$$

An alternative form is

$$x = a\cos(\omega t - \theta) \tag{2.26}$$

where the constants a, θ are related to c, d by

$$c = a\cos\theta, d = a\sin\theta \tag{2.27}$$

The constant a is called the amplitude. It is identical with the constant introduced in 2.15, and defines the extreme limits between which the particle oscillates, $x = \pm a$. The motion is a periodic oscillation, of period τ given by

$$\tau = \frac{2\pi}{\omega} \tag{2.28}$$

The frequency f is the number of oscillations per unit time,

$$f = \frac{1}{\tau} = \frac{\omega}{2\pi} \tag{2.29}$$

This discussion applies to the motion of a particle near an equilibrium point of any potential energy function. For a sufficiently small displacements, any system of this kind behaves like a simple harmonic oscillator. In particular, the frequency or period of small oscillations may always be found from the second derivative of the potential energy function at the equalbrium position.

2.3   The law of conservation energy

This law was originally derived from Newton's law for the case where the force is a function only of x. However, it has a much wider application. By introducing additional forms of energy, it has been extended far beyond the field of mechanics, to the point where it is now recognized as one of the most fundamental of all physical laws. The existence of such a law, and of the laws of conservation of momentum and angular momentum, is in fact closely related to the relativity principle.

Many non-conservative forces may be regarded as macroscopic effects of forces which are really conservative on a small scale. For example, when a particle penetrates a retarding medium, such as the atmosphere, it experiences a force which is velocity-dependent, sub-microscopic scale, we see that what happens is that the particle makes a series of collisions with the molecules of the medium. In each collision, energy is conserved, and some of the kinetic energy of the incoming particles is transferred to the molecules of the medium. In each collision, energy is conserved, and some of the kinetic energy of the incoming particle is transfered to the molecule with which it collides. By means of further collisions, this energy is gradually distributed among the surrounding molecules. The net result is to retard the incoming particles, and to increase the average energy of the molecules in the medium. This increased energy appears macroscopically as heat, and results in a rise in temperature of the medium.

For an arbitary force F, the rate of change of the kinetic energy T is given 2.3 : $\dot{T} = F\dot{x}$. Thus the increase in kinetic energy in a time interval dt, during which the particle moves a distance dx, is

$$dT = dW \tag{2.30}$$

,where dW=Fdx is called the work done by force F in the infinitesimal displacement dx. The work done is therefore a measure of the amount of energy converted to kinetic energy from other forms. In a real mechanical system, there is usually some loss of mechanical energy to heat or other forms. Correspondingly there will be dissipative forces acting on the system.

2.4   The damped Oscillator

We learned that a particle near a position of stable equilibrium under a conservative force may always be treated approximately as a simple harmonic oscillator. If there is energy loss, we must include in the equation of motion a force depending on the velocity. So long as we are concerned only with small displacements from the equalibrium position, we may treat both x and $\dot{x}$. Thus we are led to consider the damped harmonic oscillator for which

$$F = -kx - \lambda\dot{x} \tag{2.31}$$

where λ is another constant. The equation of motion now becomes

$$m\ddot{x} + \lambda\dot{x} + kx = 0 \tag{2.32}$$

Equation of this form turn up in many branches of physics. The equation 2.32 may be solved as before by looking for solutions of the form

$$x = e^{pt} \qquad (2.33)$$

Substituting in 2.32, we obtain for p the equation

$$mp^2 + \lambda p + k = 0 \qquad (2.34)$$

The solutions of this equation are

$$p = -\gamma \pm (\gamma^2 - \omega_0^2)^{1/2} \qquad (2.35)$$

where

$$\gamma = \frac{\lambda}{2m} \qquad (2.36)$$

and ω_0 is the angular frequency of the undamped oscillator,

$$\omega_0 = (k/m)^{1/2} \qquad (2.37)$$

The rate at which work is done by the force $-\lambda \dot{x}$ is $-\lambda \dot{x}^2$. If λ were negative, the particle would be gaining energy. So we shall assume that λ is positive.

2.4.1 Large damping

If λ is so large that $\gamma 2 > \omega_0^2$, then both roots for p are real and negative,

$$p = -\gamma_{\pm}, \gamma_{\pm} = \gamma \pm (\gamma^2 - \omega_0^2)^{1/2} \qquad (2.38)$$

The general solution is then

$$x = \frac{1}{2}Ae^{-\gamma_{+}t} + \frac{1}{2}Be^{-\gamma_{-}t} \tag{2.39}$$

where A and B are arbitrary constants. Hence the displacement tends exponentially to zero. For large times, the dominant term is that containing in the exponent the smaller quantity γ_{-}. Thus the characteristic time in which x is reduced by a factor $\frac{1}{e}$ is of the order of $\frac{1}{\gamma_{-}}$.

2.4.2 Small damping

Let us now consider the case when λ is small, so that $\gamma^2 < \omega_0^2$. Then the two roots for p are complex conjugates,

$$p = -\gamma \pm i\omega, \omega = (\omega_0^2 - \gamma^2)^{1/2} \tag{2.40}$$

The general solution may be written in the alternative forms

$$x = \frac{1}{2}Ae^{i\omega t - \gamma t} + \frac{1}{2}Be^{-i\omega t - \gamma t} \tag{2.41}$$

$$= Re(Ae^{i\omega t - \gamma t}) = ae^{-\gamma t}\cos(\omega t - \theta) \tag{2.42}$$

where

$$A = ae^{-i\theta}, B = ae^{i\theta} \tag{2.43}$$

From the last form of 2.42, we see that this solution represents an oscillation with exponentially decreasing amplitude $ae^{-\gamma t}$, and angular frequency ω. Note that

ω is less than the frequency ω_0 of the undamped oscillator. The time in which the amplitude is reduced by a factor $1/e$ is

$$\frac{1}{\gamma} = \frac{2m}{\lambda} \qquad (2.44)$$

This is called the relaxation time of the oscillator.

If is often convenient to introduce the quality factor, or simply Q, of the resonance, defined as the dimensionless number

$$Q = \frac{m\omega_0}{\lambda} = \frac{\omega_0}{2\gamma} \qquad (2.45)$$

If the damping is small, then Q is large. In a single oscillation period, the amplitude is reduced by the factor $e^{-\frac{2\pi\gamma}{\omega}}$, or approximately $e^{-\frac{\pi}{Q}}$. The number of periods in a relaxation time is $\frac{Q}{\pi}$.

2.4.3   Critical damping

The limiting case, $\gamma^2 = \omega_0^2$, is the case of critical damping, in which $\omega = 0$ and the two roots for p coincide. Then the solution involves only one arbitrary constant, $A + B$, and connot be the general solution. We need to find a second independent solution in this case. In fact, it is easy to verify by direct substitution that the equation 2.32 is also satisfied by the function of $te^{-\gamma t}$. Thus the general solutions is

$$x = (a + bt)e^{-\gamma t}. \qquad (2.46)$$

Critical damping is often the ideal. For example, in a measuring instrument we want to damp out the oscillation of the pointer about its correct position as quickly as possible, but too much damping would lead to a very slow response. Let us assume that k is fixed, and the amount of damping varied. When the damping is less than critical $(\gamma < \omega_0)$, the characteristic time of response is the relaxation time $\frac{1}{\gamma}$, which of course decreases as γ is increased. However, when $\gamma > \omega_0$, the characteristic time as $\frac{1}{\gamma_-}$,as we noted above. It is easy to verify that as γ increases, γ_- decreases, so that the response time $\frac{1}{\gamma_-}$ increases. Thus the shortest possible response time is obtained by choosing $\gamma = \omega_0$, that is for critical damping.

2.5   Oscillator under simple periodic force

In an isolated system, the forces are functions of position and velocity, but not explicitly of the time. However, we are often interested in the response of an oscillatory system to an applied external force, which is given as a function of the time. Then we have to consider the equation

$$m\ddot{x} + \lambda\dot{x} + kx = F(t) \tag{2.47}$$

, where $F(t)$ is the external force, Now, if $x_1(t)$ is any solution of this equation, and $x_0(t)$ is a solution of the corresponding homogeneous equation ?? for the unforced oscillator, then as one can easily check by direct substitution, $x_1(t) + x_0(t)$ will be another solution of 2.47. Hence we only need to find one particular solution of this equation. The general solution is then obtained by adding to this particular solution.

Let us consider the case where the applied force is periodic in time, with the simple form

$$F(t) = F_1 \cos \omega_1 t, \tag{2.48}$$

,where F_1 and ω_1 are constants. It is convenient to write this in the form

$$F(t) = Re(F_1 e^{i\omega_1 t}), \tag{2.49}$$

and to solve first the equation with a complex force

$$m\ddot{z} + \lambda\dot{z} + kz = F_1 e^{i\omega_1 t}. \tag{2.50}$$

Then the real part x of this solution will be a solution of the equation 2.47 with the force 2.48.

We now look for a solution of 2.50 which is periodic in time, with the same period as the applied force,

$$z = A_1 e^{i\omega_1 t} = a_1 e^{i\omega_1 t - \theta_1} \tag{2.51}$$

, where $A_1 = a_1 e^{-i\theta_1}$ is a complex constant. Substituting in 2.50 we obtain

$$(-m\omega_1^2 + i\lambda\omega_1 + k)A_1 = F_1 \tag{2.52}$$

or, dividing by $me^{-i\theta_1}$, and rearranging the terms,

$$(\omega_0^2 + 2i\gamma\omega_1 - \omega_1^2)a_1 = \frac{F_1}{m}e^{-i\theta_1}, \tag{2.53}$$

,where γ and ω_0 are defined as before by 2.36 and 2.37.

Hence equating real and imaginary parts,

$$(\omega_0^2 - \omega_1^2)a_1 = \frac{F_1}{m}\cos\theta_1, \tag{2.54}$$

$$2\gamma\omega_1 a_1 = \frac{F_1}{m}\sin\theta \tag{2.55}$$

The amplitude may be found by squaring these equations and adding, to give

$$a_1 = \frac{F_1/m}{[(\omega_0^2 - \omega_1^2)^2 + 4\gamma^2\omega_1^2]^{1/2}} \tag{2.56}$$

Dividing one equation by the other yields the phase:

$$\tan\theta_1 = \frac{2\gamma\omega_1}{\omega_0^2 - \omega_1^2} \tag{2.57}$$

If $F_1 > 0$, the correct choice between the two solutions for θ_1 differing by π is the one lying in the range $0 < \theta_1 < \pi$.

We have found a particle solution of 2.50. Its real part, that is, corresponding particular solution of our original equation, is simply

$$x = a_1\cos(\omega_1 t - \theta_1). \tag{2.58}$$

The general solution is obtained by adding to this particular solution the general solution of the corresponding homogeneous equation. In the case where the damping is less than critical ($\gamma^2 < \omega_0^2$), we obtain

$$x = a_1\cos(\omega_1 t - \theta_1) + ae^{-\gamma t}\cos(\omega t - \dot\theta). \tag{2.59}$$

Here a_1 and θ_1 are given by 2.56 and 2.57, but a and θ are arbitrary constants to be fixed by the initial conditions.

The second term in the general solution 2.59, which represents a free oscillation, dies away exponentially with time. It is therefore called the transient. After a long time, the displacement x will be given by the first term of 2.59. Thus no matter what initial conditions we choose, the oscillations are ultimately governed solely by the external force. Note that their period is the period of the applied force, not the period of the unforced oscillator.

2.5.1   Resonance

The amplitude a_1 and phase θ_1 of the forced oscillations are strongly dependent on the angular frequencies ω_0 and ω_1. In particular, if the damping γ is small, the amplitude can become very large when the frequencies are almost equal. If we fix γ and the forcing frequency ω_1, and vary the oscillator frequency ω_0, the amplitude is a maximum when $\omega_0 = \omega_1$. In this case, we say that the system is in resonance. At resonance, the amplitude is

$$a_1 = \frac{F_1}{2m\gamma\omega_1} = \frac{F_1}{\lambda\omega_1} \tag{2.60}$$

which can be very large if the damping constant λ is small. If we fix the parameters γ and ω_0 of the oscillator, and vary the forcing frequency ω_1, the maximum amplitude actually occurs for a frequency ω_1 slightly lower than ω_0, namely

$$\omega_1^2 = \omega_0^2 - 2\gamma^2 \tag{2.61}$$

However, if γ is small, this does not differ much from ω_0. Note that the natural frequency ω of the oscillator lies between this resonant frequency and the natural frequency ω_0 of the undamped oscillator.

The width of the resonance, that is the range of frequencies over which the amplitude is large, is determined by γ. For, the amplitude is reduced to $\frac{1}{\sqrt{2}}$ of its peak value when the two terms in the denominator of 2.56 become comparable in magnitude, and for small γ this occurs when $\omega_1 = \omega_0 \pm \gamma$. Therefore, γ is called the half-width of the resonance. Notice the inverse relationship between width and peak amplitude: the narrower the resonance, the higher is its peak.

The quality factor $Q = \frac{\omega_0}{2\gamma}$ provides a quantitative measure of the sharpness of the resonance peak. Indeed, the ration of the amplitude at resonance, given 2.60, to the amplitude at $\omega_1 = 0$ is precisely Q.

This phenomena of resonance occurs with any oscillatory system, and is of great practical importance. Since quite small forces can set up large oscillations if the frequencies are in resonance, great care must be taken in the design of any mechanical structure to avoid this possibility. It would be undesirable, for example, to build a ship whose natural frequency of pitching coincided with the frequency of the waves it is likely to encounter.

The constant θ_1 specified the phase relation between the applied force and the induced oscillations. If the force is slowly oscillating, ω_1 is small, and $\theta_1 \approx 0$, so that the induced oscillations are in phase with the force. In this case, the amplitude 2.56 is

$$a_1 \approx \frac{F_1}{m\omega_0^2} = \frac{F_1}{k}. \tag{2.62}$$

Thus the position s at any time t, is approximately the equalibrium position under the force $F_1 \cos \omega_1 t - kx$. At resonance, the phase shift is $\theta_1 = \frac{1}{2}\pi$, and the induced oscillations lag behind by a quarter period. For every rapidly oscillating forces, $\theta_1 \approx \pi$, and the oscillations are almost exactly out of phase. In this limiting case $a_1 \approx \frac{F_1}{m\omega_1 2}$, and the oscillations correspond to those of a free particle under the applied oscillatory force. Note that the value of the damping term γ is important only in the region near the resonance.

2.6 Collision problems

So far we have been discussing the motion of a single particle under a known external force. We now turn to the problem of an isolated system of two bodies moving under a mutual force F which is given in terms of their positions and velocities.

The equations of motion are then

$$m_1 \ddot{x}_1 = F m_2 \ddot{x}_2 = -F \tag{2.63}$$

We have already seen that these equations lead to the law of conservation of momentum,

$$p_1 + p_2 = P = constant \tag{2.64}$$

According to the relativity principle, F must be a function only of the relative distance

$$x = x_1 - x_2 \tag{2.65}$$

and the relative velocity

$$\dot{x} = \dot{x}_1 + \dot{x}_2 \tag{2.66}$$

When it is a function only of x, the force is conservative, and we can introduce a potential energy function $V(x)$ as before. The law of conservation of energy takes the form

$$T + V = E = constant \tag{2.67}$$

with

$$T = \frac{1}{2}m_1\dot{x}_1^2 + \frac{1}{2}m_2\dot{x}_2^2. \tag{2.68}$$

We are particularly interested in the collision problems, in which the force is generally small except when the bodies are very close together. The potential function is then a constant for large values of x, and rises very sharply for small values. An ideal impulsive conservative force corresponds to a potential energy function with a discontinuity or step. So long as the initial kinetic energy is less than the height of the step, the bodies will bounce off one another. From the law of conservation of energy, we can deduce that the final value of the kinetic energy, when the bodies are again far apart, is the same as the initial value before the collision. If we denote the initial velocities by u_1, u_2, and the final velocities by v_1, v_2, then

$$\frac{1}{2}m_1v_1^2 + \frac{1}{2}m_2v_2^2 = \frac{1}{2}m_1u_1^2 + \frac{1}{2}m_2u_2^2 \tag{2.69}$$

Collisions of this type, in which there is no loss of kinetic energy, are known as elastic collisions. They are typical of very hard bodies, like billiard balls.

We saw earlier that a particle which bounces back from a potential barrier emerges with velocity equal to its initial velocity. We can easily derive a similar result for the case of two-body collisions. To do this, we write the energy and momentum conservation equations in the forms

$$\frac{1}{2}m_1v_1^2 - \frac{1}{2}m_1u_1^2 = \frac{1}{2}m_2u_2^2 - \frac{1}{2}m_2v_2^2 \tag{2.70}$$

$$m_1v_1 - m_1u_1 = m_2u_2 - m_2v_2 \tag{2.71}$$

We can then divide the first of these equations by the second, and obtain

$$v_1 + u_1 = u_2 + v_2 \tag{2.72}$$

or

$$v_2 - v_1 = u_1 - u_2 \tag{2.73}$$

This shows that the relative velocity is just reversed in the collision.

Equation 2.72 and the momentum conservation equation may be solved for the final velocities in terms of the initial velocities. In particular, if the second body is initially at rest$(u_2 = 0)$, then

$$v_1 = \frac{m_1 - m_2}{m_1 + m_2} u_1 \tag{2.74}$$

$$v_2 = \frac{2m_1}{m_1 + m_2} u_1 \tag{2.75}$$

Note that if the masses are equal, then the first body is brought to rest by the collision, and its velocity transferred to the second body. If $m_1 > m_2$, the first body continues in the same direction, with reduced velocity, whereas if $m_1 < m_2$ it rebounds in the opposite direction. In the limit where m_2 is much larger than m_1, we obtain $v_1 = -u_1$, which agrees with the previous result for a particle rebounding from a fixed potential barrier.

2.6.1 Inelastic Collisions

In practice there is generally some loss of energy in a collision, for instance in the form of heat. In that case, the relative velocity will be reduced in magnitude by the collision. We define the coefficient of restitution e in a particular collision by

$$v_2 - v_1 = e(u_1 - u_2) \tag{2.76}$$

The usefulness of this quantity derives from the experimental fact that, for any two given bodies, e is approximately a constant for a wide range of velocities.

The final velocities may again be found from 2.76 for the case of elastic collision. At the other extreme, we have the case of very soft bodies which stick together on impact. Then $e = 0$, and the collision is called perfectly inelastic.

The energy loss in an inelastic collision is easily evaluated. The initial kinetic energy is $T = \frac{1}{2}m_1 u_1^2$, while the final kinetic energy is $T' = \frac{1}{2}m_1 v_1^2 + \frac{1}{2}m_2 v_2^2$. Substituting for the final velocities we find that the fractional loss of kinetic energy is

$$\frac{T - T'}{T} = \frac{(1 - e^2)m_1}{m_1 + m_2} \tag{2.77}$$

This shows that e must always be less than one, unless some energy is released, as for example by an explosion.

Chapter 3

Energy and Angular Momentum

3.1 Energy

The kinetic energy of a particle of mass m free to move in three dimensions is defined to be

$$T = \frac{1}{2}m\dot{\mathbf{r}}^2 = \frac{1}{2}m(\dot{x}^2 + \dot{y}^2 + \dot{z}^2) \tag{3.1}$$

The rate of change of the kinetic energy is therefore

$$\dot{T} = m(x\dot{x} + y\dot{y} + z\dot{z}) = m\dot{\mathbf{r}} \cdot \ddot{\mathbf{r}} = \dot{\mathbf{r}} \cdot \mathbf{F}, \tag{3.2}$$

by the equation of motion. The change in kinetic energy in a time interval dt during which the particle moves a distances $d\mathbf{r}$ is then

$$dT = dW \tag{3.3}$$

with

$$dW = \mathbf{F} \cdot d\mathbf{r} = F_x dx + F_y dy + F_z dz. \tag{3.4}$$

This is the three-dimensional expression for the work done by the force $\mathbf{F}$ in the displacement $d\mathbf{r}$. Note that it is equal to the distance travelled $|d\mathbf{r}|$ multiplied by the component of $\mathbf{F}$ in the direction of the displacement.

One might think that a conservative force in three dimensions should be defined to be a force $\mathbf{F(r)}$ depending only on the position $\mathbf{f}$ of the particle. This is not sufficient to ensure the existence of a law of conservation of energy, which is the essential feature of a conservative force. We require that

$$T + V = E = const \tag{3.5}$$

The rate of change of the potential energy is

$$\dot{V}(r) = \frac{\partial V}{\partial x}\dot{x} + \frac{\partial V}{\partial y}\dot{y} + \frac{\partial V}{\partial z}\dot{z} \tag{3.6}$$

or, in terms of the gradient of V,

$$\nabla V = \mathbf{i}\frac{\partial V}{\partial x} + \mathbf{j}\frac{\partial V}{\partial y} + \mathbf{k}\frac{\partial V}{\partial z}, \tag{3.7}$$

by

$$\dot{V} = \dot{\mathbf{r}} \cdot \nabla V. \tag{3.8}$$

Then differentiating 3.5, and with 3.2,3.8 for $\dot{T}$ and $\dot{V}$, we obtain

$$\dot{\mathbf{r}} \cdot (\mathbf{F} + \nabla V) = 0. \tag{3.9}$$

Since this must hold for any velocity of the particle, we require

$$\mathbf{F}(r) = -\nabla V(r). \tag{3.10}$$

In terms of components, it reads

$$F_x = -\frac{\partial V}{\partial x}, F_y = -\frac{\partial V}{\partial y}, F_z = -\frac{\partial V}{\partial z}, \tag{3.11}$$

For example, suppose that V has the form

$$V(r) = \frac{1}{2}k\mathbf{r}^2 = \frac{1}{2}k(x^2 + y^2 + z^2) \tag{3.12}$$

which describes a three-dimensional harmonic oscillator. Then we have

$$\mathbf{F} = (-kx, -ky, -kz) = -k\mathbf{r} \tag{3.13}$$

Any vector function $\mathbf{F}(r)$ of the form 3.10 obeys the relation

$$\nabla \times \mathbf{F} = 0 \tag{3.14}$$

that is, its curl vanishes. For instance, the z component of 3.14 is

$$\frac{\partial F_y}{\partial x} - \frac{\partial F_x}{\partial y} = 0, \tag{3.15}$$

and this is true because of 3.11 and the symmetry of the second partial derivative:

$$\frac{\partial^2 V}{\partial x \partial y} = \frac{\partial^2 V}{\partial x \partial y} \tag{3.16}$$

Thus 3.14 is a necessary condition for the force $\mathbf{F}(\mathbf{r})$ to be conservative.

3.2 Angular Momentum

The moment about the origin of a force $\mathbf{F}$ acting on a particle at position $\mathbf{r}$ is defined to be the vector product

$$\mathbf{G} = \mathbf{r} \times \mathbf{F} \tag{3.17}$$

The components of the vector $\mathbf{G}$ is the moments about the $x-, y-, z-$ axes.

$$G_x = yF_z - zF_y, G_y = zF_x - xF_z, G_x = xF_y - yF_x. \tag{3.18}$$

The direction of the vector $\mathbf{G}$ is that of the normal to the plane of $\mathbf{r}$ and $\mathbf{f}$. It may be regarded as defining the axis about which the force $\mathbf{F}$ tends to rotate the particle. The magnitude of $\mathbf{G}$ is

$$G = rF \sin \theta = bF \tag{3.19}$$

where θ is the angle between $\mathbf{r}$ and $\mathbf{F}$, and b is the perpendicular distance from the origin to the line of action of the force. Moments of forces play a particularly important role in the dynamics of rigid bodies.

Correspondingly, we define the vector angular momentum about the origin of a particle at position $\mathbf{r}$, and moving with momentum $\mathbf{p}$, to be

$$\mathbf{J} = \mathbf{r} \times \mathbf{p} = m\mathbf{r} \times \dot{\mathbf{r}} \qquad (3.20)$$

Its components, the angular momenta about the $x-, y-, z-$ axes,a re

$$J_x = m(y\dot{z} - z\dot{y}), J_y = m(z\dot{x} - z\dot{z}), J_x = m(x\dot{y} - y\dot{x}), \qquad (3.21)$$

The rate of change of the angular momentum bfJ is

$$\dot{\mathbf{J}} = m\frac{d}{dt}(\mathbf{r} \times \dot{\mathbf{r}}) = m(\dot{\mathbf{r}} \times +\mathbf{r} \times \ddot{\mathbf{r}}). \qquad (3.22)$$

The first term 3.22 is zero, because its the vector product of a vector with itself. The second term is simply $\mathbf{r} \times \mathbf{F} = \mathbf{G}$. Thus we obtain the important result that the rate of change of angular momentum is equal to the moment of the applied force.

$$\dot{\mathbf{J}} = G. \qquad (3.23)$$

This should be compared with the equation $\dot{\mathbf{p}} = \mathbf{F}$ for the rate of change of linear motion.

3.3   Central forces, Conservation of angular momentum

An external force $\mathbf{F}$ is said to be central if it is always directed towards or away from a fixed point, called the center of force. If we choose the origin to be at this center, this means that $\mathbf{F}$ to be central is that its moment about the center should vanish:

$$\mathbf{G} = \mathbf{r} \times \mathbf{F} = 0 \qquad (3.24)$$

From 3.23 it follows that when the force is central, the angular momentum is a constant:

$$\mathbf{J} = const. \qquad (3.25)$$

This is the law of conservation of angular momentum in its simplest form. It will be useful to discuss in some detail the physical significance of this law. It really contains two statements: that the direction of $\mathbf{J}$ is constant and that its magnitude is constant. The direction of $\mathbf{J}$ is that of the normal to the plane of $\mathbf{r}$ and $\mathbf{v}$. Hence the statement that this direction is fixed implies that $\mathbf{r}$ and $\mathbf{v}$ must always lie in a fixed plane. In other words, the motion of the particle is confined to the plane containing the initial position vector and velocity vector. This is obvious physically; since the force is central, it has no component perpendicular to this plane, and since the normal component of the velocity is initially zero, it must always remain zero.

To understand the meaning of the second part of the law, the constancy of the magnitude of $\mathbf{J}$, it is convenient to introduce polar coordinates r and θ in the plane of the motion. In a short time interval dt, in which the coordinates change by amounts dr and $rd\theta$, respectively. Thus the radius and transverse components of the velocity are

$$v_r = \dot{r}, v_\theta = r\dot{\theta}. \qquad (3.26)$$

The magnitude of the angular momentum is then

$$J = mrv_\theta = mr^2\dot{\theta} \tag{3.27}$$

It is easy to find a geometrical interpretation of the statement that J is a constant. We note that when the angle θ changes by an amount $d\theta$, the radius vector sweeps out an area

$$dA = \frac{1}{2}r^2 d\theta \tag{3.28}$$

Thus the rate of sweeping out area is

$$\frac{dA}{dt} = \frac{1}{2}r^2\dot{\theta} = \frac{J}{2m} = constant \tag{3.29}$$

This law is generally referred to as Kepler's second law of planetary motion, though it applies more generally to motion under any central force.

3.4 Calculus of Variations

It will be helpful to begin by discussing a very simple example. Let us ask the question; what is the shortest path between two given points in a plane? Or course we know the answer already, but the method we shall use to derive it can also be applied to less trivial examples - for instance, to find the shortest path between two points on a curved surface.

Suppose the two points are (x_0, y_0) and (x_1, y_1). Any curve joining them is represented by an equation

$$y = y(x) \tag{3.30}$$

such that the function $y(x)$ satisfies the boundary conditions

$$y(x_0) = y_0, \, y(x_1) = y_1. \tag{3.31}$$

Consider two neighbouring points on this curve. The distance dl between them is given by

$$dl = (dx^2 + dy^2)^{1/2} = (1 + y'^2)^{1/2} dx \tag{3.32}$$

where $y' = \frac{dy}{dx}$. Thus the total length of the curve is

$$l = \int_{x_0}^{x_1} (1 + y'^2)^{1/2} dx \tag{3.33}$$

Then the problem is to find that function $y(x)$, subject to the conditions 9.5, which will make this integral a minimum.

This problem differs from the usual kind of minimum-value problem in that what we have to vary is not a single variable or set of variables, but a function $y(x)$. However, we can still apply the same criterion: when the integral has a minimum value, it must be unchanged to first order by making a small variation in the function $y(x)$.

More generally, we may be interested in finding the stationary values of an integral of the form

$$I = \int_{x_0}^{x_1} f(y, y') dx, \tag{3.34}$$

where $f(y, y')$ is a specified function of y and its first derivative. We shall solve this general problem, and then apply the result to the integral 9.7. Consider a small variation $\delta y(x)$ in the function $y(x)$, subject to the condition that the values of y at the end-points are unchanged :

$$\delta y(x_0) = 0, \delta y(x_1) = 0. \tag{3.35}$$

To first order, the variation in $f(y, y')$ is

$$\delta f = \frac{\partial f}{\partial y} + \frac{\partial f}{\partial y'}\delta y', \tag{3.36}$$

where

$$\delta y' = \frac{d}{dx}\delta y. \tag{3.37}$$

Thus the variation of the integral I is

$$\delta I = \int_{x_0}^{x_1} [\frac{\partial f}{\partial y}\delta y + \frac{\partial f}{\partial y'}\frac{d}{dx}\delta y]dx. \tag{3.38}$$

In the second term, we may integrate by parts. The integrated term, namely

$$[\frac{\partial f}{\partial y'}\delta y]_{x_0}^{x_1} \tag{3.39}$$

vanished at the limits because of the conditions. Hence we obtain

$$\delta I = \int_{x_0}^{x_1} [\frac{\partial f}{\partial y} - \frac{d}{dx}(\frac{\partial f}{\partial y'})]\delta y(x)dx. \tag{3.40}$$

In order that I should be stationary, this variation δI must vanish for an arbitrary small variation $\delta y(x)$. This is only possible if the integrand vanished identically. Thus we require

$$\frac{\partial f}{\partial y} - \frac{d}{dx}\left(\frac{\partial f}{\partial y'}\right) = 0. \tag{3.41}$$

This is known as the Euler-Lagrange equation. It is general a second order differential equation for the function $y(x)$, whose solution contains two arbitrary constants that may be determined from the known values of y at x_0 and x_1.

Now we can solve the problem. In that case, comparing 9.7 and **??**, we have to choose

$$f = (1 + y'^2)^{1/2} \tag{3.42}$$

and therefore

$$\frac{\partial f}{\partial y} = 0, \frac{\partial f}{\partial y'} = \frac{y'}{(1 + y'^2)^{1/2}}. \tag{3.43}$$

Thus the Euler-Lagrange equation reads

$$\frac{d}{dx}\left[\frac{y'}{(1 + y'^2)^{1/2}}\right] = 0. \tag{3.44}$$

This equation states that the expression inside the bracket is a constant, and hence that y' is a constant. Its solutions are therefore the straight lines.

$$y = ax + b \tag{3.45}$$

Thus we have proved that the shortest path between two points is a straight line.

So far, we have used x as the independent variable, but in the applications we consider later we shall be concerned instead with functions of the time t. It is easy to generalize the discussion to the case of function f of n variables $q_2, \cdots, q_n$, and their time derivatives $\dot{q}_1, \dot{q}_2, \cdots, \dot{q}_n$. In order that the integral

$$I = \int_{t_0}^{t_1} f(q_1, \cdots, q_n, \dot{q}_1, \cdots, \dot{q}_n)dt \qquad (3.46)$$

be stationary, it must be unchanged to first order by a variation in any one of the functions $q_i(t)$, subject to the conditions $\delta q_i(t_0) = \delta q_i(t_1) = 0$. Thus we require the n Euler-Lagrange equations

$$\frac{\partial f}{\partial q_i} - \frac{d}{dt}(\frac{\partial f}{\partial \dot{q}_i}) = 0, \qquad (3.47)$$

These n second-order differential equation determine the n functions $q_i(t)$ to within $2n$ arbitrary constants of integration.

Chapter 4

Central forces

4.1 The conservation laws

We consider the general case of a central conservative force. It corresponds to a potential energy function $V(r)$ depending only on r. There are two conservation laws, one for energy, and one for angular momentum.

$$\frac{1}{2}m\dot{\mathbf{r}}^2 + V(r) = E = const, \, m\mathbf{r} \times \dot{\mathbf{r}} = \mathbf{J} = const. \tag{4.1}$$

The second of these laws implies that the motion is confined to a plane, so that the problem is effectively a two-dimensional one. Introducing polar coordinates r, θ in this plane, we may write the two conservative laws in the form

$$\frac{1}{2}m(\dot{r}^2 + r^2\dot{\theta}^2) + V(r) = E, \, mr^2\dot{\theta} = J. \tag{4.2}$$

A great deal of information can be obtained about the motion directly from these equations, without actually solving them to find r and θ as function of the time. We note that θ may be eliminated to yield an equation involving only r and $\dot{r}$,

$$\frac{1}{2}m\dot{r}^2 + \frac{J^2}{2mr^2} + V(r) = E. \tag{4.3}$$

We shall call this the radial energy equation. For a given value of J, it has precisely the same form as the one-dimensional energy equation with a potential energy function

$$U(r) = \frac{j^2}{2mr^2} + V(r). \tag{4.4}$$

It is easy to understand the physical significance of the extra term $\frac{J^2}{2mr^2}$ in this effective potential energy. It corresponds to a force $\frac{j^2}{mr^3}$. This is precisely the centrifugal force $mr\dot{\theta}^2$ expressed in terms of the constant J rather than the variable $\dot{\theta}$.

We can use the radial energy equation. Since $\dot{r}^2$ is positive, the motion is limited to the range of values of r for which

$$U(r) = \frac{J^2}{2mr^2} + V(r) \leq E. \tag{4.5}$$

The maximum and minimum radial distances are given by values of r for which the equality holds.

As an example, let us consider again the case of the isotropic oscillator, for which $V(r) = \frac{1}{2}kr^2$. The corresponding to a position of stable equilibrium in the one-dimensional at

$$r = (\frac{j^2}{mk})^{1/4}. \tag{4.6}$$

When the value of E is equal to the minimum value of $U, \dot{r}$ must always be zero, and r is fixed at the position of the minimum. In this case, the particle must move in a circle around the origin. It is interesting to note that we could also obtain 4.6 by equating the attractive force kr to the centrifugal force in the circular orbit, $\frac{J^2}{mr^3}$. For any larger value of E, the motion is confined to the region

$$b \leq r \leq a \qquad (4.7)$$

between two limiting values of r, given by the solutions of the equation 4.5. If the particle is initially at a distance r_0 from the origin, and moving with velocity v_0 in a direction making an angle α with the radial direction, then the values of E and J are

$$E = \frac{1}{2}mv_0^2 + \frac{1}{2}kr_0^2, J = mr_0v_0 \sin \alpha \qquad (4.8)$$

Thus the equation for a or b becomes

$$r^4 - (r_0^2 + \frac{m}{k}v_0^2)^2 + \frac{m}{k}r_0^2v_0^2 \sin^2 \alpha = 0. \qquad (4.9)$$

4.2   The inverse square law

We now consider a force

$$\mathbf{F} = \frac{k}{r^2}\hat{r} \qquad (4.10)$$

where k is a constant. The corresponding potential energy function is

$$V(r) = \frac{k}{r} \tag{4.11}$$

The constant k may be either positive or negative; in the first case, the force is repulsive, and in the second, attractive.

The radial energy equation for this case is

$$\frac{1}{2}m\dot{r}^2 + \frac{J^2}{2mr^2} + \frac{k}{r} = E. \tag{4.12}$$

It corresponds to the 'effective potential energy function'

$$U(r) = \frac{J^2}{2mr^2} + \frac{k}{r} \tag{4.13}$$

4.2.1 Repulsive case

We suppose first that $k > 0$. Then $U(r)$ decreases monotonically from $+\infty$ at $r = 0$ to 0 at $r = \infty$. Thus is has no minima, and circular motion is impossible, as is physically obvious. For any positive value of E, there is a minimum value of r, r_1 say, which is the unique positive root of the equation $U(r) = E$, but no maximum value. If the radial velocity is initially inward, the particle must follow an orbit on which r decreases to r_1, and then increases again without limit. As is well known, and as we shall show in the next section, the orbit is actually a hyperbola.

As an example, let us calculate the distance of closest approach for a charged particle of charge q moving in the field of a fixed point charge q'. We suppose that initially the particle is approaching the center of force with velocity v along a path which, if continued in a straight line, would pass the center at a distance b. This distance b is known as the impact parameter, and will appear frequently in our future work. Since the particle is initially at a great distance, its initial potential energy is negligible, and

$$E = \frac{1}{2}mv^2 \tag{4.14}$$

Moreover, since the component of $\mathbf{r}$ perpendicular to $\mathbf{v}$ is b, the angular momentum is

$$J = mbv \tag{4.15}$$

The distance of closest approach r_1 is obtained by substituting these values, and $k = \frac{qq'}{4\pi\epsilon}$, in the radial equation, and setting $\dot{r} = 0$.

This yields

$$r_1^2 - 2ar_1 - b^2 = 0 \tag{4.16}$$

where

$$a = \frac{qq'}{4\pi\epsilon mv^2} \tag{4.17}$$

The required solution is the positive root

$$r_1 = a + (a^2 + b^2)^{1/2} \tag{4.18}$$

4.2.2 Attractive case

We now suppose that $k < 0$. It will be useful to define a quantity l, with the dimensions of length, by

$$l = \frac{J^2}{m|k|} \tag{4.19}$$

The the 'effective potential energy function' is

$$U(r) = |k|(\frac{l}{2r^2} - \frac{1}{r}). \tag{4.20}$$

Evidently, $U(\frac{1}{2}l) = 0$, and $u(r)$ has a minimum at $r = l$, with minimum value $U(l) = -\frac{|k|}{2l}$.

Here different types of motion are possible according to the value of E. We may distinguish four cases:

$$(a)E = -\frac{|k|}{2l} \tag{4.21}$$

This is the minimum value of U. Hence $\dot{r}$ must always be zero, and the particle must move in a circle of radius l. Since the potential energy is $V = -\frac{|k|}{l}$, the kinetic energy $T = E - V$ is

$$T = \frac{1}{2}mv^2 = \frac{|k|}{2l} \tag{4.22}$$

From this equation, we can deduce the orbital velocity v. Note the interesting result that for a circular orbit the potential energy is always twice as large in magnitude as the kinetic energy.

$$(b) - \frac{|k|}{2l} < E < 0. \tag{4.23}$$

The radial distance in this case is limited between a minimum distance r_1 and a maximum distance r_2, so the motion must be periodic. As we shall see, the orbit is in fact an ellipse.

$$(c)E = 0 \tag{4.24}$$

In this case, there is a minimum distance $r_1 = \frac{1}{2}l$, but the maximum distance r_2 is infinite. Thus the particle has just enough energy to escape to infinity, with kinetic energy tending to zero at large distances. The orbit will be shown to be a parabola.

$$(d)E > 0. \tag{4.25}$$

In this case, there is a minimum distance again but no maximum distance. However, the particle can escape to infinity with non-zero limiting velocity. The orbit is in fact a hyperbola.

4.2.3 Escape velocity

As an example, we shall consider a projectile launched from the surface of the earth, with velocity v at an angle α to the vertical. The energy and angular momentum are

$$E = \frac{1}{2}mv^2 - \frac{GMm}{R}, J = mRv \sin \alpha, \tag{4.26}$$

where G is the gravitational constant. To express the energy in terms of more familiar quantities, we note that the gravitational force on a particle at the earth's surface is

$$mg = \frac{GMm}{R^2} \tag{4.27}$$

from which we obtain the useful result

$$GM = R^2 g \tag{4.28}$$

Thus,

$$E = \frac{1}{2}mv^2 - Rgm \tag{4.29}$$

The projectile will escape to infinity provided that $E \geq 0$; or that its velocity v exceeds the escape velocity

$$v_e = (2Rg)^{1/2} \tag{4.30}$$

Note that this condition is independent of the angle of projection α. Using the values $R = 6370km, g = 9.81\text{ms}^{-2}$, we find for the escape velocity from the earth

$$v_e = 11.2kms^{-1} \tag{4.31}$$

If the projectile is launched with a velocity less than the escape velocity, it will reach some maximum height and then fall back.

4.3 Orbits

We now turn to the problem of determining the orbit of a particle moving under a central conservative force. This can be done by eliminating the time from the two conservation equations 4.3 to obtain an equation relating r and θ. The simplest way of doing this is to work not with r itself, but with the variable $u = \frac{1}{r}$, and look for an equation determining u as a function of θ. Now

$$\frac{du}{d\theta} = -\frac{1}{r^2} + \frac{dr}{d\theta} \tag{4.32}$$

whence

$$\dot{r} = \frac{dr}{d\theta}\dot{\theta} = -r^2\dot{\theta}\frac{du}{d\theta} = -\frac{J}{m}\frac{du}{d\theta} \tag{4.33}$$

Thus substituting for $\dot{r}$ in the radial energy equation **??**, we obtain

$$\frac{J^2}{2m}(\frac{du}{d\theta})^2 + \frac{J^2}{2m}u^2 + V = E \tag{4.34}$$

in which of course V is to be regarded as a function of $\frac{1}{u}$. This equation can be integrated to give the equation of the orbit.

We shall consider explicitly only the case of the inverse square law, for which $V = ku$. We shall treat both cases $k > 0$ and $k < 0$ together, by writing $V = \pm|k|u$. It will be useful, as in the preceding section, to define

$$l = \frac{J^2}{m|k|} \tag{4.35}$$

Then multiplying 4.34, we obtain

$$l(\frac{du}{d\theta})^2 + lu^2 \pm 2u = \frac{2E}{|k|} \tag{4.36}$$

where the upper sign refers to the repulsive case, $k > 0$, and the lower to the attractive case, $k < 0$. To solve this equation, we multiply by l and add 1 to both sides to complete the square. Then, if we introduce the new variable

$$z = lu \pm 1, \frac{dz}{d\theta} = l\frac{du}{d\theta} \tag{4.37}$$

we may write the equation as

$$(\frac{dz}{d\theta})^2 + z^2 = \frac{2El}{|k|} + 1 = e^2 \tag{4.38}$$

Note that since the left side of the equation is a sum of squares, it can have a solution only when the right side is also positive, in agreement with our earlier result that the minimum value of E is $-\frac{|k|}{2l}$. The general solution of this equation is

$$z = lu \pm 1 = e\cos(\theta - \theta_0) \tag{4.39}$$

where θ_0 is an arbitrary constant of integration. Thus replacing $u = 1/r$ and multiplying by r , we find that the orbit equation is, in the repulsive case,

$$r[e\cos(\theta - \theta_0) - 1] = l, \tag{4.40}$$

and, in the attractive case,

$$r[e\cos(\theta - \theta_0) + 1] = l. \tag{4.41}$$

These are the polar equations of conic section, referred to a focus as origin. The constant e, the eccentricity, determines the shape of the orbit; l, called the semi-latus rectum, determines its scale; and θ_0 its orientation relative to the coordinate axes. In the repulsive case, e must be greater than unity, and therefore $E > 0$, since otherwise the square bracket in 4.40 is always negative. In the attractive case $e = 0$ when E has its minimum value $-\frac{|k|}{2l}$; the orbit is then the circle $r = l$. So long as $e < 1$, or $E < 0$, the square bracket in 4.41 is always positive, and there is a value of r for every value of θ. Thus the orbit is closed. When $e \geq 1$, or $E \geq 0$, however, r can become infinite when the square bracket vanishes.

Note that r takes its minimum value when $\theta = \theta_0$. Thus θ_0 specifies the direction of the point of closest approach. In the attractive case, the constant l also has a simple geometrical interpretation. It is the radial distance at right angles to this direction; that is, $r = l$, when $\theta = \theta_0 \pm \frac{\pi}{2}$.

4.3.1 Elliptic orbits $(E < 0, e < 1)$

For most applications, it is best to use the orbit equation directly in its polar form. It is, however, straightforward to rewrite it in the more familiar Cartesian form. If we choose the axes so that $\theta_0 = 0$, we obtain after some algebra

$$\frac{(x + ae)^2}{b^2} + \frac{y^2}{b^2} = 1, \tag{4.42}$$

where

$$a = \frac{l}{1 - e^2} = \frac{|k|}{2|E|} \tag{4.43}$$

and

$$b^2 = al = \frac{J^2}{2m|E|} \tag{4.44}$$

This is the equation of an ellipse with center at $(-ae, 0)$ and semi-axes a and b. It is useful to note that the semi-major axis a is fixed by the value of the energy, while the semi-latus rectum l is fixed byu the angular momentum.

The time taken by the particle to traverse any part of its orbit may be found from the relation between angular momentum and rate of sweeping out area,

$$\frac{dA}{dt} = \frac{J}{2m} \tag{4.45}$$

All we have to do is to evaluate the are swept out by the radius vector, and multiply by $\frac{2m}{J}$. In particular, since the area of the ellipse is $A = \pi ab$, the orbital period is $\tau = \frac{2m\pi ab}{J}$. Thus

$$\left(\frac{\tau}{2\pi}\right)^2 = \frac{m^2 a^2 b^2}{J^2} = \frac{m^3 a^3 l}{m|k|l} \tag{4.46}$$

and thus

$$\left(\frac{\tau}{2\pi}\right)^2 = \frac{m}{|k|} a^3 \tag{4.47}$$

For a planet or satellite orbiting round a central body of mass M,

$$\left(\frac{\tau}{2\pi}\right)^2 = \frac{a^3}{GM} \tag{4.48}$$

This yields Kepler's third law of planetary motion:the square of the orbital period is proportional to the cube of the semi-major axis.

4.3.2   Hyperbolar orbits $(E > 0, e < 1)$

For both the attractive and repulsive cases, the Cartesian equation of the orbit is

$$\frac{(x - ae)^2}{a^2} - \frac{y^2}{b^2} = 1. \tag{4.49}$$

where

$$a = \frac{l}{e^2 - 1} = \frac{|k|}{2E} \tag{4.50}$$

This is the equation of a hyperbola with center at $(ae, 0)$ the semi-axes a and b. One branch of the hyperbola corresponds to the orbit in the attractive case and the other to that in the repulsive case. As before, a is determined by the energy, and l by the angular momentum. Note that 4.10 and 4.11 , the semi-minor axis b is identical with the impact parameter introduced earlier.

The directions in which r becomes infinite are, in the repulsive case, $\theta = \pm \cos^-(1/e)$, and in the attractive case, $\theta = \pi \pm \cos^-(1/e)$. In both cases, the angle through which the particle is deflected from its origin line of motion is

$$\Theta = \pi - 2 \cos^{-1}(1/e). \tag{4.51}$$

This angle Θ is called the scattering angle. It will be useful to find the relation between this angle, the impact parameter b and the limiting velocity v. Since $E = \frac{1}{2}mv^2$, we have from 4.49, $a = \frac{|k|}{mv^2}$. But also, from 4.49 and 4.50,

$$b^2 = a^2(e^2 - 1) = a^2[\sec^2 \frac{1}{2}(\pi - \Theta) - 1] \tag{4.52}$$

Thus we obtain

$$b = \frac{|k|}{mv^2} \cot \frac{1}{2}\Theta \qquad (4.53)$$

4.4 Scattering Cross section

One of the most important ways of obtaining information about the structure of small bodies is to bombard them with particle and measure the number of particles scattered particles in various directions. The angular distribution of scattered particles will depend on the shape of the target, and on the nature of the forces between the particles and the target. To be able to interpret the results of such an experiment, we must know how to calculate the expected angular distribution when the forces are given.

We shall consider first a particularly simple case. We suppose that the target is a fixed, hard sphere of radius R, and that a uniform, parallel beam of particles impinges on it. Let the particle flux in the beam, that is the number of particles crossing unit area normal to the beam direction per unit time, be f. Then the number of particles which strikes the target in unit time is

$$w = f\sigma \qquad (4.54)$$

where σ is the cross-sectional area presented by the target, namely

$$\sigma = \pi R^2 \qquad (4.55)$$

Now let us consider one of these particles. We suppose that it impinges on the target with velocity v and impact parameter b. Then it will hit the target at an angle α to the normal given by

$$b = R \sin \alpha \tag{4.56}$$

The force on the particle is an impulsive central conservative force, corresponding to a potential energy function $V(r)$ which is zero for $r > R$, and rises very sharply in the neighbourhood of $r = R$. Thus the kinetic energy and angular momentum must be the same before and after the collision. We shall take the positive z direction $(\theta = 0)$ to be the direction of motion of the incoming particles. Then by the axial symmetry of the problem, the particle must move in a plane $\phi =$ constant. From energy conservation, its velocity must be the same in magnitude before and after the collision. Then, from angular momentum conservation it follows that the particle will bounce off the sphere at an angle to the normal equal to the incident angel α. Thus the particle is deflected through an dangle $\theta = \pi - 2\alpha$. Thus the particle is deflected through an angle $\theta = \pi - 2\alpha$, reflected to the impact parameter by

$$b = R \cos \frac{1}{2}\theta \tag{4.57}$$

We can not calculate the number of particles scattered in a direction specified by the polar angles θ, ϕ, within an angular range $d\theta, d\phi$. The particles scattered through angles between θ and $\theta + d\theta$ are those which came in with impact parameters between b and $b + db$, where

$$db = -\frac{1}{2}R\sin\frac{1}{2}\theta d\theta \tag{4.58}$$

Consider now a cross-section of the incoming beam. The particles we are interested in are those which cross a small region of area

$$d\sigma = b|db|d\phi \tag{4.59}$$

Inserting the values of b and db, we find

$$d\sigma = \frac{1}{4}R^2\sin\theta d\theta d\phi \tag{4.60}$$

The rate at which particles cross this area, and therefore the rate at which they emerge in the given angular range, is

$$dw = fd\sigma \tag{4.61}$$

In order to measure this rate, we may place a small detector at a large distance from the target in the specified direction. We therefore want to express our result in terms of the cross-sectional area dA of the detector, and its distance L from the target.

Now we can see that the element of area on a sphere of radius L is

$$dA = Ld\theta \times L\sin\theta d\phi \tag{4.62}$$

We define the solid angle subtended at the origin by the area dA to be

$$d\Omega = \sin\theta d\theta d\phi \tag{4.63}$$

so that

$$dA = L^2 d\Omega \tag{4.64}$$

The solid angle is measured in steradians. It plays the same role for a sphere as does the angle in radians for a circle; equation 4.64 is the analogue of the equation $ds = Ld\theta$ for a circle of radius L. Just as the total angle subtended by an entire circle is 2π, so the total solid angle subtended by an entire sphere is

$$\int\int d\Omega = \frac{1}{L^2}\int\int dA = \int_0^\pi \sin\theta d\theta \int_0^{2\pi} d\phi = 4\pi. \tag{4.65}$$

The important quantity is not the cross-sectional area $d\sigma$ itself but the ratio $\frac{d\sigma}{d\Omega}$, which is called the differential cross-section.

4.5   Rutherford Scattering

In this section we discuss a problem which was of crucial important in obtaining an understanding of the structure of the atom. In 1911 Rutheford bombarded atoms with α-particles. Because these particles are much heavier than electrons, they are deflected only very slightly by the electrons in the atom, and can therefore be used to study the heavy atomic nucleus. From observations of the angular distribution of the between α-particles, Rutherford was able to show that the law of force small distances. Thus he concluded that the positive charge is concentrated in a very small nuclear volume rather than being spread out over the volume of the atom.

We shall calculate the differential cross-section for the scattering of a particle of charge q and mass m by a fixed point charge q'. The impact parameter b is related to the scattering angle θ by 4.53, i.e.

$$b = a \cot \frac{1}{2}\theta, a = \frac{qq'}{4\pi\epsilon_0 mv^2} \tag{4.66}$$

Thus

$$db = -\frac{ad\theta}{2\sin^2\frac{1}{2}\theta} \tag{4.67}$$

so that, substituting in 4.59, we obtain

$$d\sigma = \frac{a^2 \cos\frac{1}{2}\theta d\theta d\phi}{2\sin^3\frac{1}{2}\theta} \tag{4.68}$$

Dividing by the solid angle, we find for the differential cross-section,

$$\frac{d\sigma}{d\Omega} = \frac{a^2}{4\sin^4\frac{1}{2}\theta}, a = \frac{qq'}{4\pi\epsilon_0 mv^2}. \tag{4.69}$$

This is the Rutherford scattering cross-section.

We note that, in contrast to the differential cross-section for scattering by a hard sphere, this cross-section is strongly dependent both on the velocity of the incoming particle and on the scattering angle. It also increases rapidly with increasing charge. For scattering of an α-particle on a nucleus of atomic number Z, $qq' = 2Ze^2$, where e is the electronic charge. Thus we expect the number of particles scattered to increase like Z^2 with increasing atomic number.

We saw that the minimum distance of approach is given by 4.18. Thus to investigate the structure of the atom at small distances, we must use high-velocity particles, for which a is small, and examine the parameter. The cross-section is large for small values of the scattering angle, but physically it is the large-angle scattering which is of interest. For the fact that particles can be scattered through large angles is an indication that there are very strong forces acting at short distances. If the positive nuclear charge were spread out a large volume, the force would be the inverse- square-law force only down to a distance equal to the radius of the charge distribution. Beyond this point, it would decrease as we go to within this distances. Consequently, the particles which penetrate to within this distance would experience a weaker force than the inverse square law predicts, and would be scattered through smaller angles.

A peculiar feature of the differential cross-section is that the corresponding total cross-section is infinite. This is a consequence of the infinite range of the Coulomb force. However far away from the nucleus a particle may be, it still experiences some force, and is scattered through a non-zero angle. Thus the total number of particles scattered through any angle, however small, in indeed infinite. We can easily calculate the number of particles scattered through any angle greater than some lower limit θ_0. There are the particles which had impact parameter b less than $b_0 = a \cot \frac{1}{2}\theta_0$. The corresponding cross-section is therefore

$$\sigma(\theta > \theta_0) = \pi b_0^2 = \pi a^2 \cot^2 \frac{1}{2}\theta. \tag{4.70}$$

Chapter 5

Rotating Frames

We have always used inertial frames, in which the laws of motion take on the simple form expressed in Newton's laws. There are a number of problems which can most easily be solved by using a non-inertial frame. For example, when discussing the motion of a particle near the earth's surface, it is often convenient to use a frame which is rigidly fixed to the earth, and rotates with it. In this chapter, we shall find the equation of motion with respect to such a frame, and discuss some applications of them.

5.1   Angular velocity

Let us consider a solid body which is rotating with constant angular velocity ω about a fixed axis. let $\mathbf{n}$ be a unit vector along the axis, whose direction is defined by the right-hand rule; it is the direction in which a right-hand-thread screw would move when turned in the direction of the rotation. Then we define the vector angular velocity ω to be a vector of magnitude ω in the direction of $\mathbf{n}, \omega = \omega\mathbf{n}$. Clearly angular velocity, like angular momentum, is an axial vector.

For example, for the earth, ω is a vector pointing along the polar axis towards the north pole. Its magnitude is equal to 2π divided by the length of the sidereal day(the rotation period with respect to the fixed stars, which is less than that with respect to the sun by one part in 365), that is

$$\omega = \frac{2\pi}{86164}s^{-1} = 7.29 \times 10^{-5}s^{-1} \tag{5.1}$$

If we take the origin to line on the axis of rotation, then the velocity of a point of the body at position $\mathbf{r}$ is given by the simple formula

$$\mathbf{v} = \omega \times \mathbf{r}. \tag{5.2}$$

To prove this, we note that the point moves with angular velocity ω round a circle of radius $\rho = r\sin\theta$. Thus its speed is

$$v = \omega\rho = \omega r\sin\theta = |\omega \times \mathbf{r}|. \tag{5.3}$$

Moreover the direction of $\mathbf{v}$ is that of $\omega \times \mathbf{r}$; for clearly, $\mathbf{v}$ is perpendicular to the plane containing ω and $\mathbf{r}$, and it is easy to see that its sense is correctly given by the right-hand rule.

It is not necessary that $\mathbf{r}$ should be the position vector of a point of the rotating body. If $\mathbf{a}$ is any vector fixed in the rotating body, then by the same argument

$$\frac{d\mathbf{a}}{dt} = \omega \times \mathbf{a}. \tag{5.4}$$

In particular, if $\mathbf{i}, \mathbf{j}, \mathbf{k}$ are unit vectors fixed in the body, then

$$\frac{d\mathbf{i}}{dt} = \omega \times \mathbf{i}, \frac{d\mathbf{j}}{dt} = \omega \times \mathbf{j}, \frac{d\mathbf{k}}{dt} = \omega \times \mathbf{k}. \tag{5.5}$$

For example, if ω is in the $\mathbf{k}$ direction, then

$$\frac{d\mathbf{i}}{dt} = \omega \mathbf{j}, \frac{d\mathbf{j}}{dt} = -\omega \mathbf{i}, \frac{d\mathbf{k}}{dt} = 0. \tag{5.6}$$

Now consider a vector $\mathbf{a}$ specified with respect to the rotating axises $\mathbf{i}, \mathbf{j}, \mathbf{k}$, by the components a_x, a_y, a_z, so that

$$\mathbf{a} = a_x \mathbf{i} + a_y \mathbf{j} + a_z \mathbf{k}. \tag{5.7}$$

We must distinguish two kinds of 'rates of change' , and it will be convenient to suspend for the moment the convention whereby $d\mathbf{a}/dt$ the rate of change of $\mathbf{a}$ as measured by an inertial observer at rest relative to the origin, and by $\dot{\mathbf{a}}$ the rate of change as measured by an observer rotating with the solid body. We wish to find the relation between these two rates of change.

Now, although our two observers differ about the rate of change of a vector, they will always agree about the rate of change of the components a_x, a_y, a_z. Hence we can write

$$\frac{da_x}{dt} = \dot{a}_x, \frac{da_y}{dt} = \dot{a}_y, \frac{da_z}{dt} = \dot{a}_z. \tag{5.8}$$

According to the observer on the rotating body, the rate of change of $\mathbf{a}$ is fully describes by the rates of change of its components, so that

$$\dot{\mathbf{a}} = \dot{a}_x \mathbf{i} + \dot{a}_y \mathbf{j} + \dot{a}_z \mathbf{k}. \tag{5.9}$$

However, to the inertial observer, the axes $\mathbf{i}, \mathbf{j}, \mathbf{k}$ are themselves changing with time, so that

$$\frac{d\mathbf{a}}{dt} = (\frac{da_x}{dt}\mathbf{i} + \frac{da_y}{dt}\mathbf{j} + \frac{da_z}{dt}\mathbf{k}) + (a_x\frac{\mathbf{i}}{dt} + a_y\frac{\mathbf{j}}{dt} + a_z\frac{\mathbf{k}}{dt}) \tag{5.10}$$

$$= (\dot{a}_x\mathbf{i} + \dot{a}_y\mathbf{j} + \dot{a}_z\mathbf{k}) + \omega \times (a_x\mathbf{i} + a_y\mathbf{j} + a_z\mathbf{k}), \tag{5.11}$$

Thus finally we obtain

$$\frac{d\mathbf{a}}{dt} = \dot{\mathbf{a}} + \omega \times \mathbf{a}. \tag{5.12}$$

Applied to a position vector $\mathbf{r}$, this result is almost obvious, for it states that the velocity with respect to an inertial observer is the sum of the velocity $\dot{\mathbf{r}}$ with respect to an inertial observer is the sum of the velocity $\dot{\mathbf{r}}$ with respect to the rotating frame, and the velocity $\omega \times \mathbf{r}$ of a particle at $\mathbf{r}$ rotating with the body.

We shall frequently encounter in our later work equations similar to 5.4: $\frac{d\mathbf{a}}{dt} = \omega \times \mathbf{a}$. It is important to realize that one can reverse the argument that led up to it. If $\mathbf{a}$ satisfies this equation, then it must be a vector of constant length rotating with angular velocity ω about the direction of ω. For, if we introduce a frame of reference rotating with angular velocity ω, then according to 5.12, $\dot{\mathbf{a}} = 0$, so that $\mathbf{a}$ is fixed in the rotating frame.

5.2   Apprarent gravity

We can use the formular 5.12 twice to obtain the relation between the absolute acceleration $d^2\mathbf{r}/dt^2$ of a particle and its acceleration $\ddot{\mathbf{r}}$ relative to a rotating frame. The velocity of the particle relative to an inertial observer is

$$\mathbf{v} = \frac{d\mathbf{r}}{dt} = \dot{\mathbf{r}} + \omega \times \mathbf{r}. \tag{5.13}$$

Applying the same formula to the rate of change of $\mathbf{v}$, we have

$$\frac{d^2\mathbf{r}}{dt^2} = \frac{d\mathbf{v}}{dt} = \dot{\mathbf{v}} + \omega \times \mathbf{v}. \tag{5.14}$$

But from 5.13,

$$\dot{\mathbf{v}} = \ddot{\mathbf{r}} = \omega \times \dot{\mathbf{r}} \tag{5.15}$$

and

$$\omega \times \mathbf{v} = \omega \times \dot{\mathbf{r}} + \omega \times (\omega \times \mathbf{r}). \tag{5.16}$$

Hence substituting in 5.14, we obtain

$$\frac{d^2\mathbf{r}}{dt^2} = \ddot{\mathbf{r}} + 2\omega \times \dot{\mathbf{r}} + \omega \times (\omega \times \mathbf{r}). \tag{5.17}$$

The second term on this right is called the "Coriolis accerelation", and the third term the "centripetal acceleration". The latter is directed inwards the axis, and perpendicular to it, and can be written in the form

$$\omega \times (\omega \times \mathbf{r}) = (\omega \cdot \mathbf{r})\omega - \omega^2 \mathbf{r}. \tag{5.18}$$

The most important application of 5.17 is to a particle moving near the surface of the earth. For a particle moving under gravity, and under an additional mechanical force $\mathbf{F}$, the equation of motion is

$$m\frac{d^2\mathbf{r}}{dt^2} = m\mathbf{g} + \mathbf{F} \tag{5.19}$$

where $\mathbf{g}$ is a vector of magnitude g pointing downward. Using 5.17, and moving all the terms except the relative acceleration to the right side of the equation, we can write it as

$$m\ddot{\mathbf{r}} = m\mathbf{g} + \mathbf{F} - 2m\omega \times \dot{\mathbf{r}} - m\omega \times (\omega \times \mathbf{r}) \tag{5.20}$$

The last two terms on the right are apparent forces, which arise because of the non-inertial nature of the frame of reference. The third term is the Coriolis force and the last term is the centrifugal force.

This force is a slowly varying function of position, proportional , like the gravitational force, to the mass of the particle. When we make a laboratory measurement of the acceleration due to gravity, what we actually measure is not $\mathbf{g}$ but

$$\mathbf{g}^* = \mathbf{g} - \omega \times (\omega \times \mathbf{r}). \tag{5.21}$$

In particular, a plumb line does not point diredtly towards the earth's center , but is swing through a small angle by the centrifugal force. Let us consider a point in colatitude ($\frac{1}{2}\pi-$latitude)θ. Then

$$|\omega \times (\omega \times \mathbf{r}| = \omega|\omega \times \mathbf{r}| = \omega^2 r \sin \theta \tag{5.22}$$

Thus the horizontal and vertical components of $\mathbf{g}^*$ are

$$g_h^* = \omega^2 r \sin \theta \cos \theta, g_v^* = g - \omega^2 r \sin^2 \theta. \tag{5.23}$$

The magnitude of the centrifugal force may be found by inserting the value for $\omega = \frac{2\pi}{86164} s^{-1} = 7.29 \times 10^{-5} s^{-1}$, and for r the mean radius of the earth, 6370km. Then we find

$$\omega^2 r = 34 mms^{-2} \tag{5.24}$$

Since $\omega^2 r \ll g$, and the angle α between the apparent and true verticals is approximately

$$\alpha \approx \frac{g_h^*}{g_v^*} \approx \frac{\omega^2 r}{g} \sin \theta \cos \theta \tag{5.25}$$

The maximum value occurs at $\theta = 45$ degree.

At the pole, there is no centrifugal force, and $g^* = g$. On the equator, $g^* = g - \omega^2 r$. Thus we might expect the measured value of the acceleration due to the gravity to be larger at the pole by $34 mms^{-2}$. the actual measured difference is somewhat larger than this,

$$\Delta g^* = g_{pole}^* - g_{eq}^* = 52 mms^{-2}. \tag{5.26}$$

This discrepancy arises from the fact that the earth is not a perfect sphere, but more nearly spheroidal in shape, flattened at the poles. Thus the gravitational acceleration g, even excluding the centrifugal term, is itself larger at the pole than on the equator. These two effects are not really independent, for the flattening of the earth is a consequence of its rotation.

5.3   Coriolis force

The Coriolis force $-2m\omega \times \dot{\mathbf{r}}$ is an apparent velocity-dependent force arising from the earth's rotation. To understand its physical origin, it may be helpful to consider a flat rotating disc. Suppose that a particle moves across the disc under no forces, so that an inertial observer sees it moving diametrically across in a straight line. Then, because the disc is rotating, an observer on the disc will see the particle crossing successive radii, following a curved track. If he is unaware that the disc is rotating, he will ascribe this curvature to a force acting on the particle at right angles to its velocity. This is the Coriolis force.

We shall examine some effects of the Coriolis force which observable in the laboratory. Let us consider a particle moving near the surface of the earth in colatitude θ. We shall suppose that the distance through which it moves is sufficiently small for both the gravitational and centrifugal forces to be effectively constant. Then we may combine them in a constant effective gravitational acceleration $\mathbf{g}^{*}$. For a convenience, we shall drop the star, and write this constant simply as $\mathbf{g}$. The equation of motion is then

$$m\ddot{\mathbf{r}} = m\mathbf{g} + \mathbf{F} - 2m\omega \times \dot{\mathbf{r}}. \tag{5.27}$$

Let us choose our axes to that $\mathbf{i}$ is east, $\mathbf{j}$ north, and $\mathbf{k}$ up.

Here 'up' means opposite to $\mathbf{g}^*$, and θ is the angle between this direction and the earth's axis. The angular velocity vector then has the components.

$$\omega = (0, \omega \sin \theta, \omega \cos \theta). \tag{5.28}$$

Hence the Coriolis force is

$$-2m\omega \times \dot{\mathbf{r}} = 2m\omega(\dot{y}\cos\theta - \dot{z}\sin\theta, -\dot{x}\cos\theta, \dot{x}\sin\theta). \tag{5.29}$$

5.3.1 Freely falling body

As the first example, let's consider that a particle is dropped from rest at a height h above the ground. Neglecting the Coriolis force, its motion is described by

$$x = 0, y = 0, z = h - \frac{1}{2}gt^2. \tag{5.30}$$

We shall calculate the effect of the Corilos force to first order; that is, we neglect terms of order ω^2. Since the Coriolis force contains a factor of ω, this means that we may substitute for $\dot{\mathbf{r}}$ the zero-order value

$$\dot{x} = 0, \dot{y} = 0, \dot{z} = -gt. \tag{5.31}$$

Then the equation of motion read

$$m\ddot{x} = 2m\omega gt \sin\theta, m\ddot{y} = 0, m\ddot{z} = -mg. \tag{5.32}$$

The solution with appropriate initial condition is

$$x = \frac{1}{3}\omega g t^3 \sin\theta, y = 0, z = h - \frac{1}{2}g t^2. \tag{5.33}$$

The particle will hit the ground, $z = 0$, at a point east of that vertically below its point of release, at a distance

$$x = \frac{1}{3}\omega(\frac{8h^3}{g})^{1/2}\sin\theta \tag{5.34}$$

For example, if a particle is dropped form a height of 100m in latitude 45 degree, then the deviation is about 16mm.

5.3.2 Foucault's pendulum

Another way of observing the effect of the Coriolis force is to use Foucault's pendulum. This is simply an ordinary pendulum free to swing in any direction, and carefully arranged to be perfectly symmetric, so that its periods of oscillation in all directions are precisely equal. The pendulum should be long and fairly heavy, so that it will go on swinging freely for several hours at least, despite the resistance of the air. If the amplitude amplitude is small, the pendulum equation is simply the equation of two-dimensional simple harmonic motion. The vertical component of the Coriolis force is negligible, for it is mearely a small correction to g, whose sign alternates on each half-period. The important components are the horizontal ones. For small amplitude, the velocity of the pendulum bob is almost horizontal, so that $\dot{z} \approx 0$. Thus the equation of motion for the x and y coordinates are

$$\ddot{x} = -\frac{g}{l}x + 2\omega\dot{y}\cos\theta, \ddot{y} = -\frac{g}{l}y - 2\omega\dot{x}\cos\theta. \tag{5.35}$$

or in vector notation

$$\ddot{\mathbf{r}} = -\frac{g}{l}\mathbf{r} - 2\omega\cos\theta\mathbf{k} \times \dot{\mathbf{r}}. \tag{5.36}$$

Let us first suppose that the pendulum is at the north pole. Then it is clear from the equation of motion in a non-rotating frame that it must swing in a fixed direction in space, while the earth rotates beneath it. Thus, relative to the earth, its oscillation plane must rotate around the vertical with angular velocity $-\omega$. Now at any other latitude, the only difference 5.36 is that in place of the angular velocity ω we have only its vertical component $\omega\cos\theta\mathbf{k}$. Hence we should expect that the pendulum rotates with angular velocity $-\Omega = -\omega\cos\theta$ around the vertical. In effect, we regard the earth's surface in colatitude θ as rotating about the vertical with angular velocity $\omega\cos\theta$.

We can verify this conclusion by obtaining an explicit solution of equation 5.35. A way of doing this is to combine the two equation by writing $z = x + iy$. Then $\dot{y} - i\dot{x} = -i\dot{z}$, so that the equations become

$$\ddot{z} + 2i\Omega\dot{z} + \omega_0^2 = 0. \tag{5.37}$$

where $\Omega = \omega\cos\theta$ and $\omega_0^2 = \frac{g}{l}$. We look for solutions of the form $z = Ae^{pt}$, and the equation for p is

$$p^2 + 2i\Omega p + \omega_0^2 = 0. \tag{5.38}$$

The roots of this equation are $p = -i\Omega \pm i\omega_1$, where $\omega_1^2 = \omega_0^2 + \Omega^2$. Hence the general solution of the equation for z is

$$z = Ae^{-i(\Omega - \omega_1)t} + Be^{-i(\Omega - \omega_1)t}. \qquad (5.39)$$

In particular, if we set $A = B = \frac{1}{2}a$, we obtain a solution

$$z = ae^{-i\Omega t} \cos \omega_1 t, \qquad (5.40)$$

or, in terms of x and y,

$$z = a \cos \Omega t \cos \omega_1 t, y = -a \sin \Omega t \cos \omega_1 t. \qquad (5.41)$$

Initially the oscillation is in the x direction. As time progresses, the amplitude $a \cos \Omega t$ of the x coordinate decreases, which that of the y coordinate, $-a \sin \Omega t$, grows. The solution represents an oscillation of amplitude a in a plane rotating with angular velocity $-\Omega$.

At the pole, the plane of oscillation makes a complete revolution in just 24 hours. At any other latitude the period is greater than this, and is in fact $\frac{2\pi}{\omega \cos \theta}$. In latitude 45 degree, it is about 34 hours, while on the equator it is infinite.

Chapter 6

The few body problem

6.1 Center of mass and relative coordinates

We denote the positions and masses of the two particles by $\mathbf{r}_1, \mathbf{r}_2$ and m_1, m_2. If the force on the first particle due to the second is $\mathbf{F}$, then by Newton's third law, that on the second due to the first is $-\mathbf{F}$. Thus in a uniform gravitational field $\mathbf{g}$ the equation of motion are

$$m_1\ddot{\mathbf{r}}_1 = m_1\mathbf{g} + \mathbf{F}, m_2\ddot{\mathbf{r}}_2 = m_2\mathbf{g} - \mathbf{F}. \qquad (6.1)$$

It is convenient to introduce new variables in place of $\mathbf{r}_1$ and $\mathbf{r}_2$. We define the position of the center of mass.

$$\mathbf{R} = \frac{m_1\mathbf{r}_1 + m_2\mathbf{r}_2}{m_1 + m_2}. \qquad (6.2)$$

and the relative position

$$\mathbf{r} = \mathbf{r}_1 - \mathbf{r}_2. \tag{6.3}$$

From 6.1 we get the equation of motion for $\mathbf{R}$ by adding,

$$M\ddot{\mathbf{R}} = M\mathbf{g}, M = m_1 + m_2, \tag{6.4}$$

and that for $\mathbf{r}$ by dividing by the masses and then subtracting,

$$\mu\ddot{\mathbf{r}} = \mathbf{F}, \mu = \frac{m_1 m_2}{m_1 + m_2}. \tag{6.5}$$

These are two equations are now completely separate. Equation 6.4 shows that the center of mass' moves with uniform acceleration $\mathbf{g}$. In the case $\mathbf{g} = 0$. It is equivalent to the law of conservation of momentum.

$$M\dot{\mathbf{R}} = m_1\dot{\mathbf{r}}_1 + m_2\dot{\mathbf{r}}_2 = \mathbf{P} = const. \tag{6.6}$$

The equation of motion for the relative position is identical with the equation for a single particle of mass μ moving under the force $\mathbf{R}$ and $\mathbf{r}$ as function of time, we can obtain the positions of the particles by solving the simultaneous equations 6.2 and 6.3.

$$\mathbf{r}_1 = \mathbf{R} + \frac{m_2}{M}\mathbf{r}, \mathbf{r}_2 = \mathbf{R} - \frac{m_1}{M}\mathbf{r}. \tag{6.7}$$

The separation between center of mass and relative motion extends to the expressions for the total angular momentum and kinetic energy. We have

$$\mathbf{J} = m_1\mathbf{r}_1 \times \dot{\mathbf{r}}_1 + m_2\mathbf{r}_2 \times \dot{\mathbf{r}}_2 \tag{6.8}$$

$$= m_1(\mathbf{R} + \frac{m_2}{M}\mathbf{r}) \times (\dot{\mathbf{R}} + \frac{m_2}{M}\dot{\mathbf{r}}) + m_2(\mathbf{R} - \frac{m_1}{M}\mathbf{r}) \times (\dot{\mathbf{R}} - \frac{m_1}{M}\dot{\mathbf{r}}). \tag{6.9}$$

It is easy to see that the cross terms between $\mathbf{R}$ and $\mathbf{r}$ cancel, so that we are left with

$$\mathbf{J} = m\mathbf{R} \times \dot{\mathbf{R}} + \mu\mathbf{r} \times \dot{\mathbf{r}}. \tag{6.10}$$

Similarly, substituting 6.7 into

$$T = \frac{1}{2}m_1\dot{\mathbf{r}}_1^2 + \frac{1}{2}m_2\dot{\mathbf{r}}_2^2. \tag{6.11}$$

We find after a little algebra,

$$T = \frac{1}{2}M\dot{\mathbf{R}}^2 + \frac{1}{2}\mu\dot{\mathbf{r}}^2. \tag{6.12}$$

6.2   The center of mass frame

It is often convenient to describe the motion of the system in terms of a frame of reference in which the center of mass is at rest at the origin. This is called the center-of-mass(CM) frame. We shall denote quantities referred to it by an asterisk.

The relative position $\mathbf{r}$ is of course independent of the choice of origin, so that setting $\mathbf{R}^* = 0$ in 6.7 we find

$$\mathbf{r}_1^* = \frac{m_2}{M}\mathbf{r}, \mathbf{r}_2^* = \frac{m_1}{M}\mathbf{r}. \tag{6.13}$$

In this frame, the momenta of the two particles are equal and opposite,

$$m_1\dot{\mathbf{r}}_1^* = -m_2\dot{\mathbf{r}}_2^* = \mu\dot{\mathbf{r}} = \mathbf{p}^*, \tag{6.14}$$

As we shall see explicitly later, it is often convenient to solve a problem first in the CM frame. To find the solution in some other frame, we then need the relations between the momenta in the two frames. Let us consider a frame in which the center of mass is moving with velocity $\dot{\mathbf{R}}$. Then the velocities of the two particles are

$$\dot{\mathbf{r}}_1 = \dot{\mathbf{R}} + \dot{\mathbf{r}}_1^*, \dot{\mathbf{r}}_2 = \dot{\mathbf{R}} + \dot{\mathbf{r}}_2^*. \tag{6.15}$$

Hence by 6.14 their momenta are

$$\mathbf{p}_1 = m_1\dot{\mathbf{r}}_1 = m_1\dot{\mathbf{R}} + \mathbf{p}^*, \mathbf{p}_2 = m_2\dot{\mathbf{r}}_2 = m_2\dot{\mathbf{R}} - \mathbf{p}^*. \tag{6.16}$$

From 6.10 and 6.12 it follows that the total angular momentum and kinetic energy in the CM frame are

$$\mathbf{J}^* = \mu\mathbf{r} \times \dot{\mathbf{r}} = \mathbf{r} \times \mathbf{p}^*, \tag{6.17}$$

$$T^* = \frac{1}{2}\mu\dot{\mathbf{r}}^2 = \frac{\mathbf{p}^{*2}}{2\mu}. \tag{6.18}$$

Thus in any other frame we can write

$$\mathbf{P} = M\dot{\mathbf{R}}, \mathbf{J} = M\mathbf{R} \times \dot{\mathbf{R}} + \mathbf{J}^*, T = \frac{1}{2}M\dot{\mathbf{R}}^2 + T^*. \qquad (6.19)$$

To obtain the values in any frame from those in the CM frame, we have only to add the contribution of a particle of mass M located at the center of mass $\mathbf{R}$.

6.3   Elastic Collisions

A collision between two particles is called elastic if there is no loss of kinetic energy in the collision: that is, if the total kinetic energy after the collision is the same as that before it. Such collisions are typical of very hard bodies, like billiard balls.

It is easy to describe such a collision in the CM frame. The particles must approach each other with equal and opposite momenta, $\mathbf{p}^*$ and $-\mathbf{p}^*$, and recede after the collision again with equal and opposite momenta, $\mathbf{q}^*$ and $-\mathbf{q}^*$. Thus each particle is scattered through the same angle θ^*. Since the collision is elastic, we have

$$T^* = \frac{\mathbf{p}^{*2}}{2\mu} = \frac{\mathbf{q}^{*2}}{2\mu}. \qquad (6.20)$$

Thus the magnitudes of the momenta before and after the collision are the same,

$$p^* = q^* \qquad (6.21)$$

In practice, most experiments are performed with one particle initially at rest in the laboratory. To interpret such an experiment, we therefore need to use the laboratory frame, in which the momentum of particle 2 before the collision is zero, $\mathbf{p}_2 = 0$.

We shall denote the Lab momentum of the incoming particle by $\mathbf{p}_1$, the momenta after the collision by $\mathbf{q}_1$ and $\mathbf{q}_2$, and the angles of scattering and recoil by θ and α. We could work out the relations between these quantities by using the conservation laws of momentum and energy directly in the Lab frame, but it is actually simpler to relate them to the CM quantities.

Since $\mathbf{p}_2 = 0$, we have

$$\dot{\mathbf{R}} = \frac{1}{m_2}\mathbf{p}^*. \tag{6.22}$$

and also

$$\mathbf{p}_1 = \frac{m_1}{m_2}\mathbf{p}^* + \mathbf{p}^* = \frac{M}{m_2}\mathbf{p}^*. \tag{6.23}$$

The momenta after the collision are given by 6.16 but with $\mathbf{q}^*$ in place of $\mathbf{p}^*$. They are

$$\mathbf{q}_1 = \frac{m_1}{m_2}\mathbf{p}^* + \mathbf{q}^*, \mathbf{q}_2 = \mathbf{p}^* - \mathbf{q}^*. \tag{6.24}$$

By 6.21 the vectors $\mathbf{p}^*, \mathbf{q}^*, \mathbf{p}_2$ form an isosceles triangle. Thus the recoil angle α and the recoil momentum $\mathbf{q}_2$ are given in terms of CM quantities by

$$\alpha = \frac{1}{2}(\pi - \theta^*), q_2 = 2p^* \sin\frac{1}{2}\theta^*. \tag{6.25}$$

The Lab kinetic energy transferred to the target particle is therefore

$$T_2 = \frac{q_2^2}{2m_2} = \frac{2p^{*2}}{m_2}\sin^2\frac{1}{2}\theta^*. \tag{6.26}$$

On the other hand, the total kinetic energy in the Lab is just the kinetic energy of the incoming particle,

$$T = \frac{p_1^2}{2m_1} = \frac{M^2 p^{*2}}{2m_1 m_2^2} \tag{6.27}$$

The interesting quantity is the fraction of the total kinetic energy which is transferred. This is

$$\frac{T_2}{T} = \frac{4m_1 m_2}{M^2} \sin^2 \frac{1}{2}\theta^*. \tag{6.28}$$

The maximum possible kinetic energy transfer occurs for a head-on collision($\theta^* = \pi$), and is $\frac{4m_1 m_2}{(m_1+m_2)^2}$. Clearly, this can be close to unity only if m_1 and m_2 are comparable in magnitude. If the incoming particle is very light, it bounces off the target with little loss of energy; if it is very heavy, it is hardly deflected at all from its original trajectory, and again loses little of its energy. For example, in a proton-α-particle collision ($\frac{m_1}{m_2}$ = 4 or 1/4), the maximum fractional energy transfer is 64 percent; in a proton-electron collision it is about 0.2 percent.

Another important relation is that between the Lab and CM scattering angles. It is easy to prove by elementary trigometry that

$$\tan\theta = \frac{\sin\theta^*}{(m_1/m_2) + \cos\theta^*}. \tag{6.29}$$

Note that the relation is independent of the momenta of the two particles, and depends only on their mass ratio.

6.4 Many-body systems

Any material object may be regarded as composed of a large number of small particles, small enough to be treated as essentially point-like, but still large enough to obey the laws of classical rather than quantum mechanics. These particles interact in complicated ways with each other and with the environment. However if we are interested only in the motion of the object as a whole, many of these details are irrelevant.

Here we consider a general system of N particles labelled by an index $i = 1, 2, \cdots, N$, interacting through two-body forces and subjected also to external forces due to bodies outside the system. We denote the force on the ith particle due to the jth particle by $\mathbf{F}_{ij}$, and the external force on the ith particle by $\mathbf{F}_i$. Thus the equations of motion are

$$m_i \ddot{\mathbf{r}}_i = \mathbf{F}_{i1} + \mathbf{F}_{i2} + \cdots + \mathbf{F}_{iN} + \mathbf{F}_i = \sum_j \mathbf{F}_{ij} + \mathbf{F}_i. \qquad (6.30)$$

Here the sum is over all particles of the system. Of course there is no force on the ith particle due to itself, and so $\mathbf{F}_{ii} = 0$. This sum in this case is really over the other $N - 1$ particles.

6.5 Momentum, center-of-mass motion

The position $\mathbf{R}$ of the center of mass is defined by

$$\mathbf{R} = \frac{1}{M} \sum_i m_i \mathbf{r}_i, \, M = \sum_i m_i. \tag{6.31}$$

The total momentum is

$$\mathbf{p} = \sum_i m_i \dot{\mathbf{r}}_i = M\dot{\mathbf{R}}. \tag{6.32}$$

It is equal to the momentum of a particle of mass M located at the center of mass.

The rate of change of momentum is

$$\dot{\mathbf{P}} = \sum_i \sum_j \mathbf{F}_{ij} + \sum_i \mathbf{F}_i. \tag{6.33}$$

Now the two-body forces $\mathbf{F}_{ij}$ must satisfy Newton's third law,

$$\mathbf{F}_{ji} = -\mathbf{F}_{ij} \tag{6.34}$$

Thus, for every term $\mathbf{F}_{ij}$ in the double sum in 6.33, there is an equal and opposite term $\mathbf{F}_{ij}$. The terms therefore cancel in pairs, and the double sum is zero.

Hence we obtain the important result that the rate of change of momentum is equal to the sum of the external forces alone

$$\dot{\mathbf{P}} = M\ddot{\mathbf{R}} = \sum_i \mathbf{F}_i. \tag{6.35}$$

In the special case of an isolated system of particles, acted on by no external forces, this yields the law of conservation of momentum

$$\mathbf{P} = M\dot{\mathbf{R}} = const. \tag{6.36}$$

In this case, the center of mass moves with uniform velocity.

Let us now regard our system of particles as forming a composite body. If the body is isolated, then according to Newton's first law it moves with uniform velocity. Thus we see that to maintain this law for composite bodies, we should define the position of such a body to mean the position of its center of mass. Moreover, with this definition, Newton's second law is just 6.35 , provided that we interpret the force on the body in the obvious way to mean the sum of the forces on all its constituent particles, and the mass as the sum of their masses. It is also clear that if Newton's third law applies to each pair of particles from two composite bodies, then it will applies to each pair of particles from two composite bodies, then it will apply to the bodies as a whole.

Thus, with suitable interpretation of the concepts involved, Newton's three laws may be applied to composite bodies as well as to point particles. It follows that, so long as we are interested only in the motion of a body as a whole, we may replace it by a particle of mass M located at the center of mass. This is a result of the greatest importance, for it allows us to apply our earlier discussion of particle motion to real physical bodies. We have implicitly assumed it in many of our applications: for example, we treated the planets as point particles in discussing their orbital motion.

6.6 Angular momenta: Central internal forces

The total angular momentum of our system of particles is

$$\mathbf{J} = \sum_i m_i \mathbf{r}_i \times \dot{\mathbf{r}}_i \tag{6.37}$$

The rate of change of $\mathbf{J}$ is

$$\dot{\mathbf{J}} = \sum_i m_i \mathbf{r}_i \times \ddot{\mathbf{r}}_i = \Sigma_i \sum_j \mathbf{r} \times \mathbf{F}_{ij} + \sum_i \mathbf{r}_i \times \mathbf{F}_i. \tag{6.38}$$

Now let us examine the contribution to 6.38 from j the internal force between a particular pair of particles, 1 and 2. For simplicity, let us write $\mathbf{F} = \mathbf{F}_{12} = -\mathbf{F}_{21}$ and $\mathbf{r} = \mathbf{r}_1 + \mathbf{r}_2$. The contribution consists of two terms, $\mathbf{r}_1 \times \mathbf{F}_{12}$ and $\mathbf{r}_2 \times \mathbf{F}_{21}$. Thus it is

$$\mathbf{r}_1 \times \mathbf{F} - \mathbf{r} \times \mathbf{F} = \mathbf{r} \times \mathbf{F}. \tag{6.39}$$

This contribution will be zero if $\mathbf{F}$ is a central force, parallel to $\mathbf{r}$. Let us for the moment simply assume that all the internal forces are central. Then the terms in the double sum of 6.38 will cancel in pairs, just as they did in the evaluation of the rate of change of total momentum. The total moment of all the internal forces will be zero. So, when the internal forces are central, the rate of change of angular momentum is equal to the sum of the moments of the external forces,

$$\dot{\mathbf{J}} = \sum_i \mathbf{r}_i \times \mathbf{F}_i. \tag{6.40}$$

In particular, for an isolated system, we have the law of conservation of angular momentum,

$$\mathbf{J} = const. \tag{6.41}$$

More generally this is true if all the external forces are directed towards, or away from the origin.

Although we have only shown that 6.40 and 6.41 hold in the special case of central internal forces, they are actually of much more general validity. It is certainly not true that all the internal forces in real solids, liquids or gases are central. Many are notably the electromagnetic force between moving charges. It is better to regard as a basics postulate of the dynamics of composite bodies, justified by the fact that the predictions derived from it agree with observation.

It is often convenient to separate the contributions to $\mathbf{J}$ from the center of mass motion and the relative motion. We define the position $\mathbf{r}_i^*$ of the particles relative to the center of mass by

$$\mathbf{r}_i + \mathbf{R} + \mathbf{r}_i^*. \tag{6.42}$$

Clearly, the position of the center of mass relative to itself is the zero vector, so that

$$\sum_i m_i \mathbf{r}_i^* = 0. \tag{6.43}$$

Now substituting 6.42 into 6.37, we obtain

$$\mathbf{J} = (\sum_i m_i)\mathbf{R} \times \dot{\mathbf{R}} + (\sum_i m_i \mathbf{r}_i^*) \times \dot{\mathbf{R}} + \mathbf{R} \times (\sum_i m_i \mathbf{r}_i^*) + \Sigma_i m_i \mathbf{r}_i^* \times \dot{\mathbf{r}}_i^*. \quad (6.44)$$

The second and third terms vanish in virture of 6.44. Thus we can write

$$\mathbf{J} = M\mathbf{R} \times \dot{\mathbf{R}} + \mathbf{J}^*. \qquad (6.45)$$

where $\mathbf{J}^*$, the angular momentum about the center of mass, is

$$\mathbf{J}^* = \sum_i m_i \mathbf{r}_i^* \times \dot{\mathbf{r}}_i^*. \qquad (6.46)$$

It is easy to find the rate of change of $\mathbf{J}^*$.

$$\frac{d}{dt}(M\mathbf{R} \times \dot{\mathbf{R}}) = M\mathbf{R} \times \ddot{\mathbf{R}} = \mathbf{R} \times \sum_i \mathbf{F}_i. \qquad (6.47)$$

Hence subtracting from 6.40 and using 6.42 we obtain

$$\mathbf{J}^* = \sum_i \mathbf{r}_i^* \times \mathbf{F}_i. \qquad (6.48)$$

Thus the rate of change of $\mathbf{J}^*$ is equal to the sum of the moments of the external forces about the center of mass. This is remarkable result. Because it must be remembered that the center of mass is not in general moving uniformly. In general, we may take moments about the origin of any inertial frame, but it would be quite wrong to take moments about an accelerated point. Only in the special case where the point is the center of mass are we allowed to do this.

This result means that in discussing the rotational motion of a body we can ignore the motion of the center of mass, and treat is as though it were fixed. It is particularly important in the case of rigid bodies.

For an isolated system, $\mathbf{J}^*$ as well as $\mathbf{J}$ is a constant. More generally,$\mathbf{J}^*$ is constant if the external forces have zero total moment about the center of mass. For example, for a system of particles in a uniform gravitational filed, the resultant force acts at the center of mass, and $\mathbf{J}^*$ is therefore a constant.

Chapter 7

Rigid Bodies

The principal characteristics of a solid is its rigidity. Under normal circumstances, its size and shape vary only slightly under stress, changes in temperature and the like. Thus it is natural to consider the idealization of a perfectly rigid body, whose size and shape are permanently fixed. Such a body may be characterized by the requirement that the distance between any two points of the body remains fixed. Here we shall be concerned with the mechanics of rigid bodies.

7.1   The fundamentals

It will be convenient to simplify the notation of the previous chapter by omitting the particle label i form sums over all particles in the rigid body. Thus we shall write

$$\mathbf{P} = \sum m\dot{\mathbf{r}}, \mathbf{J} = \sum m\mathbf{r} \times \dot{\mathbf{r}}. \tag{7.1}$$

The motion of the center of mass of the body is completely specified by

$$\dot{\mathbf{P}} = M\ddot{\mathbf{R}} = \sum \mathbf{F}. \tag{7.2}$$

Our main interest will be focused on the rotational motion of the body. For the moment let us assume, that the internal forces are central. Then, according to 6.40

$$\dot{\mathbf{J}} = \sum \mathbf{r} \times \mathbf{F}. \tag{7.3}$$

We shall see that these two equations are sufficient to determine the motion completely.

One very important implication of 7.1 and 7.2 should be noted. For a rigid body at rest, $\ddot{\mathbf{r}} = 0$ for every particle, and thus $\mathbf{P}$ and $\mathbf{J}$ both vanish. Clearly the body can only remain at rest if the right hand sides of both 7.1 and 7.2 vanish. That is if the sum of the forces and the sum of their moments are both zero. In fact, as we shall see, this is not only a necessary but also a sufficient condition for equilibrium.

Under the same assumption of central internal forces, the internal forces in a rigid body do not work, so that

$$\dot{T} = \sum \dot{\mathbf{r}} \cdot \mathbf{F}. \tag{7.4}$$

This might appear at first sight to be a third independent equation. However we shall see later that it is actually a consequence of the other two. It is particularly useful in the case when the external forces are conservative, since it leads to the conservation law

$$T + V = E = const. \tag{7.5}$$

where V is the external potential energy.

The assumption that the internal forces are central is much stronger than it need be. Indeed, it could not be justified from our knowledge of the internal forces in real solids, which are certainly not exclusively central, and in any case cannot adequately be described by classical mechanics. All we actually requires is the validity of equation 7.1 and 7.2, and it is better to regard these as basic assumptions of rigid body dynamics, whose justification lies in the fact that their consequences agree with observation.

7.2   Rotation about a fixed axis

Let us now apply these basic equations to a rigid body which is free to rotate only about a fixed axis, which for simplicity we take to be the z axis. We also choose the position of the origin on this axis so that the z coordinate of the center of mass is 0.

In cylindrical polars, the z and ρ coordinates of every point are fixed, while the ϕ coordinate varies according to $\dot{\phi} = \omega$, the angular velocity of the body. We examine first the component of angular momentum about the axis of rotation. It is

$$J_z = \sum m\rho v_\phi = I\omega. \tag{7.6}$$

where

$$I = \sum m\rho^2. \tag{7.7}$$

is the moment of inertia about the z axis. Since I is obviously constant, the z component of 7.2 yields

$$\dot{J}_z = I\dot{\omega} = \sum \rho F_\phi. \tag{7.8}$$

This equation determines the rate of change of angular velocity, and may be called the equation of motion of the rotating body.

The condition for equailibrium here is simply that the right hand side of 7.8 vanishes, i.e. that the forces have no net moment about the z axis.

For example, consider a rectangular lamina, of size $a \times b$, pivoted at one corner, carrying a weight Mg and supported by a horizontal force F. The total moment is $bF - aMg$, and thus we find that for equalibrium $F = (a/b)Mg..$

The kinetic energy may also be expressed in terms of I. For

$$T = \sum \frac{1}{2}m(\rho\dot{\phi})^2 = \frac{1}{2}I\omega^2. \tag{7.9}$$

The equation 7.3 for the rate of change of kinetic energy is

$$\dot{T} = I\omega\dot{\omega} = \sum (\rho\dot{\phi})F_\phi = \omega \sum \rho F_\phi. \tag{7.10}$$

This is clearly a consequence of 7.8, and gives us no additional information.

The momentum equation 7.1 serves to determine the reaction at the axis, which has no moment about the axis, and does not appear in 7.8. Let us denote the force on the body at the axis by $\mathbf{Q}$, and separate this from the other forces on the body. Then from 7.1

$$\dot{\mathbf{P}} = M\ddot{\mathbf{r}} = \mathbf{Q} + \sum \mathbf{F}. \qquad (7.11)$$

For example, the equilibrium condition $\dot{\mathbf{P}} = 0$ immediately yields $\mathbf{Q} = (F, Mg, 0)$.

The center of mass is fixed in the body, so that $\dot{\mathbf{R}} = \omega \times \mathbf{R}$. Differentiating to find the acceleration of the center of mass, we obtain.

$$\ddot{\mathbf{R}} = \dot{\omega} \times \mathbf{R} + \omega \times \dot{\mathbf{R}} = dot\omega \times \mathbf{R} + \omega \times (\omega \times \mathbf{R}). \qquad (7.12)$$

The first term is the tangential accleration, $R\dot{\omega}$ in the ϕ direction, and the second the radial acceleration, $-\omega^2 R$ in the ρ direction. Together 7.11 and 7.12 determine $\mathbf{Q}$.

7.3 Perpendicular components of angular momentum

We now consider the remaining components of the angular momentum vector $\mathbf{J}$. We shall see that they give us further information about the reaction on the axis. In Cartesian coordinates the velocity of a point $\mathbf{r}$ of the body is given by

$$\dot{x} = -\omega y, \dot{y} = \omega x, \dot{z} = 0. \qquad (7.13)$$

Thus we find

$$J_x = \sum m(-z\dot{y}) = -\sum mxz\omega, \; J_y = \sum m(z\dot{x}) = -\sum myz\omega. \qquad (7.14)$$

We can write all three components of $\mathbf{J}$ in the form

$$J_x = I_{xz}\omega, \; J_y = I_{yz}\omega, \; J_z = I_{zz}\omega. \qquad (7.15)$$

where

$$I_{xz} = -\sum mxz, \; I_{yz} = -\sum myz, \; I_{zz} = -\sum m(x^2 + y^2). \qquad (7.16)$$

Here I_{zz} is the moment of inertial about the z-axis, previously denoted by I. The quantities I_{xz} and I_{yz} are called products of inertia.

At first sight it may seem surprising that $\mathbf{J}$ has components in direction perpendicular to ω . A simple example may help to clarify the reason for this. Consider a light rigid rod with equal masses m at its two ends, rigidly fixed at an angle θ to an axis through its mid-point. If the positions of the masses are $\mathbf{r}$ and $-\mathbf{r}$, the total angular momentum is

$$\mathbf{J} = m\mathbf{r} \times \dot{\mathbf{r}} + m(-\mathbf{r}) \times (-\dot{\mathbf{r}}) = 2m\mathbf{r} \times (\omega \times \mathbf{r}). \qquad (7.17)$$

Clearly, $\mathbf{J}$ is perpendicular to $\mathbf{r}$. When the rod is in the xz-plane, the masses are moving in the $\pm y$ directions, and there is a component of angular momentum about the x-axis as well as one about the z-axis.

In this example, the center of mass lies on the axis. Thus if no external force is applied, the total reaction on the axis is zero. There is, however, a resultant couple on the axis, which is required to balance the couple produced by the centrifugal forces. The magnitude of this couple may be determined from the remaining pair of the equation 7.2. From 7.16 unlike I_{zz}, the products of inertia are not constants.

Using 7.13, we find

$$\dot{I}_{xz} = -\omega I_{yz}, \dot{I}_{yz} = \omega I_{xz}, \dot{I}_{zz} = 0. \tag{7.18}$$

Thus, if **G** is the couple on the axis, we have

$$\dot{J}_x = I_{xz}\dot{\omega} - I_{yz}\omega^2 = G_x + \sum(yF_z - zF_y), \dot{J}_y = I_{yz}\dot{\omega} + I_{xz}\omega^2 = G_y + \sum(zF_x - xF_z) \tag{7.19}$$

If there are no external forces, then ω is constant, and **G** precisely balances the centrifugal couple. When the rod is in the xz-plane, the only non-vanishing component is the moment about the y-axis,

$$G_y = I_{xz}\omega^2 = -2mr^2\omega^2 \sin\theta \cos\theta. \tag{7.20}$$

7.4 Principal axes of inertia

We have seen that the angular momentum vector **J** is in a different direction form the angular velocity vector ω. There are special cases in which the products of inertia I_{xz} and I_{yz} vanish. Then **J** is also in the z direction. In that case, the z-axis is called a principal axis of inertia.

When a body is rotating freely about a principal axis through its center of mass, there is no resultant force or couple on the axis. The z-axis is a principal axis if the xy-plane is a plane of reflection symmetry; for then the contribution to the products of inertia I_{xz} and I_{yz} form any point (x, y, z) is exactly cancelled by that from the point $(x, y, -z)$. Similarly, it is a principal axis if it is an axis of rotational symmetry, for then the contribution from (x, y, z) is cancelled by that from $(-x, y, -z)$.

For bodies with three symmetry axes there are obviously three perpendicular principal axes, and we shall see later that this is true. In this case, it is clearly an advantage to choose these as our coordinate axes. So we shall no longer assume that the axis of rotation is the z-axis but take it to have an arbitrary inclination. Then ω has three components $(\omega_x, \omega_y, \omega_z)$.

The angular momentum vector $\mathbf{J}$ may be written in the form

$$\mathbf{J} = \sum m\mathbf{r} \times (\omega \times \mathbf{r}) = \Sigma m[r^2\omega - (\mathbf{r} \cdot \omega)\mathbf{r}]. \tag{7.21}$$

Its components are therefore linear functions of the components of ω, which we may write in matrix notation as

$$\begin{pmatrix} J_x \\ J_y \\ J_z \end{pmatrix} = \begin{pmatrix} I_{xx} & I_{xy} & I_{xz} \\ I_{yx} & I_{yy} & I_{yz} \\ I_{zx} & I_{zy} & I_{zz} \end{pmatrix} \begin{pmatrix} \omega_x \\ \omega_y \\ \omega_z \end{pmatrix}, \tag{7.22}$$

that is, $J_x = I_{xx}\omega_x + I_{xy}\omega_y + I_{xz}\omega_z$, etc. The nine elements $I_{xx}, I_{xy}, \cdots, I_{zz}$ of the 3×3 matrix may be regarded as components of a single entity I, in much the same way that the quantities J_x, J_y, J_z are regarded as components of the vector $\mathbf{J}$. The entity I is called a tensor, in this case the inertial tensor.

It is obvious from the definition 7.16 that the products of inertial satisfy relations like $I_{xy} = I_{yx}$. The 3×3 matrix is therefore unchanged by reflection in the leading diagonal. A tensor I with this property is called symmetric.

If the three coordinate axes are all axes of symmetry, then all the products of inertia vanish, and I has the diagonal form

$$\begin{pmatrix} I_{xx} & 0 & 0 \\ 0 & I_{yy} & 0 \\ 0 & 0 & I_{zz} \end{pmatrix}. \tag{7.23}$$

In this case, the relation 7.22 simplify to

$$J_x = I_{xx}\omega_x, \, J_y = I_{yy}\omega_y, \, J_z = I_{zz}\omega_z. \tag{7.24}$$

Then $\mathbf{J}$ is parallel to ω if the axis of rotation is any one of the three symmetry axes, but not in general otherwise.

It is shown that for any given symmetric tensor one can always find a set of axes with respect to which it is diagonal. Thus for any rigid body we can find three perpendicular axes through any given point which are principal axes of inertia. It will be convenient to introduce three unit vectors $\mathbf{e}_1, \mathbf{e}_2, \mathbf{e}_3$ along these axes. Then if we write

$$\omega = \omega_1 \mathbf{e}_1 + \omega_2 \mathbf{e}_2 + \omega_3 \mathbf{e}_3. \tag{7.25}$$

the components of $\mathbf{J}$ in these three directions will be obtained by multiplying the components of ω by the appropriate moments of inertia. Thus we obtain

$$\mathbf{J} = I_1\omega_1\mathbf{e}_1 + I_2\omega_2\mathbf{e}_2 + I_3\omega_3\mathbf{e}_3. \tag{7.26}$$

The three diagonal elements of the inertia tensor, I_1, I_2, I_3, are called principal moments of inertia. We shall always use a single subscript for the principal moments, to distinguish them from moments of inertia about arbitrary axes.

It is important to realize that the principal axes $\mathbf{e}_1, \mathbf{e}_2, \mathbf{e}_3$ are fixed in the body, not in space, and therefore rotate with it. It is often convenient to use these axes to define our frame of reference, particularly since the principal moments I_1, I_2, I_3 are constants. This is however a rotating frame, not an inertial one.

The kinetic energy T may also be expressed in terms of the angular velocity and the inertia tensor. We have

$$T = \sum \frac{1}{2}m\dot{\mathbf{r}}^2 = \sum \frac{1}{2}m(\omega \times \mathbf{r})^2 = \sum \frac{1}{2}m[\omega^2 r^2 - (\omega\dot{\mathbf{r}})^2], \tag{7.27}$$

by a standard formula of vector algebra.

Comparing with 7.21, we see that

$$T = \frac{1}{2}\mathbf{J} \cdot \omega. \tag{7.28}$$

Thus we get

$$T = \frac{1}{2}I_1\omega_1^2 + \frac{1}{2}I_2\omega_2^2 + \frac{1}{2}I_3\omega_3^2. \tag{7.29}$$

These equations for rotational motion may be compared with the corresponding ones for translational motion with velocity $\mathbf{v}, T = \frac{1}{2}M\mathbf{v}^2 = \frac{1}{2}\mathbf{Pv}$. The principal difference is that mass, unlike the inertia tensor, has no directional properties, so that the coefficients of v_x^2, v_y^2 and v_z^2 are all equal.

7.5 Calculation of moments of inertia

The moments and products of inertia of any body with respect to a given origin may be calculated from the definition 7.16. For continuous distributions of matter, we must replace the sums by integrals,

$$I_{xx} = \int \int \int \rho(\mathbf{r})(y^2 + z^2)d^3\mathbf{r}, I_{xy} = \int \int \int \rho(\mathbf{r})(-xy)d^3\mathbf{r}. \qquad (7.30)$$

etc., where $\rho(\mathbf{r})$ is the dencity.

7.5.1 Shift of origin

When the rigid body is pivoted so that one point is fixed, it is convenient to choose that point to be the origin. If there is no fixed point, we generally choose the origin to be at the center of mass. Thus it is useful to be able to relate the moments and products of inertia about an arbitrary origin to those about the center of mass.

As usual, we shall distinguish quantities referred to the center of mass as origin by an asterisk. To find the desired relations. we substitute in 7.16 $\mathbf{r} = \mathbf{R} + \mathbf{r}^*$, and use the relations 6.44 or,

$$\sum mx^* = \sum my^* = \sum mz^* = 0. \tag{7.31}$$

Because of these relations, the cross terms between $\mathbf{R}$ and $\mathbf{r}^*$ drop out, exactly as they did in 6.45 . For example,

$$I_{xy} = -\sum m(X + x^*)(Y + y^*) = -MXY - \sum mx^*y^*. \tag{7.32}$$

The last term is just the product of inertia I_{xy}^* referred to the center of mass at origin. Thus we obtain relations of the form

$$I_{xx} = M(Y^2 + Z^2) + I_{xx}^*, I_{xy} = -MXY + I_{xy}^*. \tag{7.33}$$

Note that the components of the inertia tensor with respect to an arbitrary origin are obtained from those with respect to the center of mass by adding the contribution of a particle of mass M at $\mathbf{R}$. Because of this result, it is only necessary, for any given body, to compute the moments and products of inertia with respect to the center of mass.

It is important to realize that the principal axes at a given origin are not necessarily parallel to those at the center of mass. If we choose the axes at the center of mass to be principal axes, then the products of inertia $I_{xy}^*, \cdots$, will be zero. but it is clear that this does not necessarily imply that $I_{xy}^*, \cdots$, are zero. In fact, this will be true only if the chosen origin lies on one of the principal axes through the center of mass, so that two of the three center-of-mass coordinates, X, Y, Z are zero.

7.6   Stability of rotation about a principal axis

The principal axes $\mathbf{e}_1, \mathbf{e}_2, \mathbf{e}_3$ rotate with the rigid body. Thus, if we wish to use directly the expression for $\mathbf{J}$ in terms of its components with respect to these axes, we have to remember that they constitute a rotating frame.

$$\frac{d\mathbf{J}}{dt} = \sum \mathbf{r} \times \mathbf{F} = \mathbf{G}. \tag{7.34}$$

The relative rate of change if

$$\dot{\mathbf{J}} = I_1\dot{\omega}_1\mathbf{e}_1 + I_2\dot{\omega}_2\mathbf{e}_2 + I_3\dot{\omega}_3\mathbf{e}_3. \tag{7.35}$$

since the principal moments of inertia are constants. The two rates of change are related by

$$\frac{d\mathbf{J}}{dt} = \dot{\mathbf{J}} + \omega \times \mathbf{J}. \tag{7.36}$$

Substituting in 7.34, we obtain

$$\dot{\mathbf{J}} + \omega \times \mathbf{J} = \mathbf{G}. \tag{7.37}$$

in terms of components,

$$I_3\dot{\omega}_3 + (I_2 - I_1)\omega_1\omega_2 = G_3. \tag{7.38}$$

and two similar equations obtained by cyclic permutation of 1,2,3.

The equation 7.38 may be solved, in principle, to give the angular velocity components as functions of time. However, when there are external forces, it is not a particularly useful equation, because these forces are usually specified in terms of their components with respect to a fixed set of axes. Even if the external force **F** is a constant, its components F_1, F_2, F_3 are variable, and depend on the unknown orientation of the body.

We shall confine our discussion to the case where there are no external forces, so that the right side of 7.38 vanishes. If the body is initially rotating about the principal axis $\mathbf{e}_3$, so that $\omega_1 = \omega_2 = 0$, then we see from 7.38 that ω_2 is a constant. The other two equations show similarly that ω_1 and ω_2 remain zero. Thus we have verified our earlier assertion that a body rotating freely about a principal axis will continue to rotate with constant angular velocity.

Now we wish to investigate the stability of this motion; that is, we ask what happens if the body is given a small displacement, so that its axis of rotation no longer precisely conincide with $\mathbf{e}_3$. Under these circumstances, ω_1 and ω_2 will be small, but not precisely zero. If the displacement is small enough , we may neglect the product $\omega_1\omega_2$, so that from 7.38 we again learn that ω_3 is a constant. The other two equations are

$$I_1\dot{\omega}_1 + (I_3 - I_2)\omega_2\omega_3 = 0, I_2\dot{\omega}_2 + (I_1 - I_3)\omega_3\omega_1 = 0. \tag{7.39}$$

We can solve these equations by looking for solutions of the form

$$\omega_1 = a_1 e^{pt}, \omega_2 = a_2 e^{pt}. \tag{7.40}$$

where a_1 and a_2 are constants.

Substituting, and eliminating the ratio $\frac{a_1}{a_2}$, we obtain

$$p^2 = \frac{(I_3 - I_2)(I_1 - I_3)}{I_1 I_2}\omega_3^2. \tag{7.41}$$

The denominator is obviously positive. Thus, if $I_3 > I_1$ and $I_3 > I_2$, or if $I_3 < I_1$ and $I_3 < I_2$, the two roots for p are pure imaginary, and we have an oscillatory solution. On the other hand, if $I_1 > I_3 > I_2$ or $I_1 < I_3 < I_2$, then the roots are real, and the values of ω_1 and ω_2 will in general increase exponentially with time.

Therefore we may conclude that the rotation about the axis $\mathbf{e}_3$ is stable if I_3 is either the largest or the smallest of the three principal moments, but not if it is the middle one. This interesting result is quite easy to verify. For example, if one throws a match-box in the air, it is not hard to get it to spin about its shortest or longest axis, but it will not spin stably about the other one.

Chapter 8

Small Oscillations and normal modes

8.1 Equations of motion for small oscillation

Now let us consider the potential energy function V. With T given by

$$\ddot{q}_\alpha = -\frac{\partial V}{\partial q_\alpha}. \tag{8.1}$$

for each $\alpha = 1, 2, \cdots, n$. Thus the condition for equalibrium is that all n partial derivatives of V should vanish at the equilibrium position. For small values of the coordinates, we can expand V in a series, just as we did for a single coordinate.

For example, for $n = 2$,

$$V = V_0 + (b_1 q_1 + b_2 q_2) + (\frac{1}{2}k_{11}q_1^2 + k_{12}q_1 q_2 + \frac{1}{2}k_{22}q_2^2) + \cdots. \tag{8.2}$$

The equalibrium conditions require that the linear terms should be zero, $b_1 =$

$b_2 = 0$. Moreover the constant V_0 is arbitrary, and may be set equal to zero. Thus the leading terms are the quadratic ones, and for small values of q_1 and q_2 we may approximate V by

$$V = \frac{1}{2}k_{11}q - 1^2 + k_{12}q_1q_2 + \frac{1}{2}k_{22}q_2^2. \tag{8.3}$$

Then the equation of motion become

$$\ddot{q}_1 = -k_{11}q_1 - k_{12}q_2, \ddot{q}_2 = -k_{21}q_1 - k_{22}q_2. \tag{8.4}$$

where for the sake of symmetry we have written $k_{12} = k_{21}$.

In the general case, V may be taken to be a homogeneous quadratic function of the coordinates, which can be written

$$V = \Sigma_{\alpha=1}^{n} \sum_{\beta=1}^{n} \frac{1}{2}k_{\alpha\beta}q_\alpha q_\beta. \tag{8.5}$$

with $k_{\alpha\beta} = k_{\beta\alpha}$.

Then the equation of motion are

$$\ddot{q}_\alpha = -\sum_{\beta}^{n} k_{\alpha\beta}q_\beta. \tag{8.6}$$

or in matrix notation

$$
\begin{pmatrix} \ddot{q}_1 \\ \ddot{q}_2 \\ \cdots \\ \ddot{q}_n \end{pmatrix} = \begin{pmatrix} k_{11} & k_{12} \cdots & k_{1n} \\ k_{21} & k_{22} \cdots & k_{2n} \\ \cdots & \cdots & \cdots & \cdots \\ k_{n1} & k_{n2} \cdots & k_{nn} \end{pmatrix} \begin{pmatrix} q_1 \\ q_2 \\ \cdots \\ q_n \end{pmatrix}, \tag{8.7}
$$

For example, in the case of the double pendulum, the potential energy is

$$
V = (M + m)gL(1 - \cos\theta) + mgl(1 - \cos\theta). \tag{8.8}
$$

For small angles, we can approximate $1 - \cos\theta$ by $\frac{1}{2}\theta^2$. Hence in terms of the orthogonal coordinates we obtain

$$
V = \frac{(M + m)g}{2L}x^2 + \frac{mg}{2l}(y - x)^2. \tag{8.9}
$$

Thus the equation of motion are

$$
\begin{pmatrix} \ddot{x} \\ \ddot{y} \end{pmatrix} = \begin{pmatrix} -\frac{M+m}{ML}g - \frac{mg}{Ml} & \frac{mg}{Ml} \\ \frac{g}{l} & -\frac{g}{l} \end{pmatrix} \begin{pmatrix} x \\ y \end{pmatrix}, \tag{8.10}
$$

Note the inequity of the coefficients corresponding to k_{12} and k_{21} of 8.4. This arises form the fact that we have not absorbed the normalization factors $M^{1/2}$ and $m^{1/2}$ into x and y.

8.2 Normal modes

The general solution of the pair of second-order differential equations 8.4 must involve four arbitrary constants, which may be fixed by the initial values of $q_1, q_2, \dot{q}_1, \dot{q}_2$.

Similarly, the general solution of 8.7 must involve $2n$ arbitrary constants. To find this general solution, we look first for solutions in which all the coordinates are oscillating with the same frequency ω,

$$q_\alpha = A_\alpha e^{i\omega t}, \tag{8.11}$$

where the A_α are complex constants. Such solutions are called normal modes of oscillation of the system.

Substituting 8.11 into 8.7, we obtain a set of n simultaneous linear equations for the n amplitudes A_α,

$$-\omega^2 A_\alpha = -\sum_{\beta}^{n} k_{\alpha\beta} A_\beta. \tag{8.12}$$

Let us consider the case $n = 2$. Then these equations are

$$\begin{pmatrix} k_{11} & k_{12} \\ k_{21} & k_{22} \end{pmatrix} \begin{pmatrix} A_1 \\ A_2 \end{pmatrix} = \omega^2 \begin{pmatrix} A_1 \\ A_2 \end{pmatrix}. \tag{8.13}$$

This is what is known as an eigenvalue equation. The values of ω^2 for which non-zero solutions exist are called the eigen values of the 2×2 matrix with elements $k_{\alpha\beta}$. The column vector formed by the A_α is an eigenvector of the matrix.

The equations 8.13 can alternatively be written as

$$\begin{pmatrix} k_{11} - \omega^2 & k_{12} \\ k_{21} & k_{22} - \omega^2 \end{pmatrix} \begin{pmatrix} A_1 \\ A_2 \end{pmatrix} = \begin{pmatrix} 0 \\ 0 \end{pmatrix}. \tag{8.14}$$

These equations have a non-zero solution if and only if the determinant of the coefficient matrix vanishes,

$$\begin{vmatrix} k_{11} - \omega^2 & k_{12} \\ k_{21} & k_{22} - \omega^2 \end{vmatrix} = (k_{11} - \omega^2)(k_{22} - \omega^2) - k_{12}^2 = 0. \tag{8.15}$$

This is called the characteristic equation for the system. It determines the frequencies ω of the normal modes, which are the square roots of the eigenvalues ω^2.

Equation 8.15 is a quadratic equation for ω^2. Its discriminant may be written $(k_{11} - k_{12})^2 + 4k_{12}^2$, which is clearly positive. Hence it always has two real roots. The condition for stability is that both roots should be positive. A negative root, say $-\gamma^2$, would yield a solution of the form

$$q_\alpha = A_\alpha e^{\gamma t} + B_\alpha e^{-\gamma t}. \tag{8.16}$$

where both the A_α and the B_α coefficients constitute eigenvectors of the 2×2 matrix, corresponding to9 the eigenvalue $-\gamma^2$. Except in the degenerate case of two equal eigenvalues, this means that the A_α and the B_α coefficients must be proportional, since the eigenvector is unique up to a factor. In general, the solution yields an exponential increase in the displacements with time.

The stability requirement that all the eigenvalues be positive is a natural generalization of the requirement, in the one-dimensional case, that the second derivative of the potential energy function be positive.

If ω^2 is chosen equal to one of the two roots of 8.15, then either of the two equations in 8.13, determines the ratio $\frac{A_1}{A_2}$. Since the coefficients are real numbers, the ratio is obviously real. This means that A_1 and A_2 have the same phase, so that

q_1 and q_2 not only oscillate with the same frequency, but actually in phase. The ratio of q_1 to q_2 remains fixed throughout the motion.

There remains in A_1 and A_2 a common arbitrary complex factor, which serves to fix the overall amplitude and phase of the normal mode solution. Thus each normal mode solution contains two arbitrary real constants. Since the equations of motion 8.4 are linear , any linear superposition of solutions is again a solution. Hence the general solution in simply a superposition of the two normal mode solutions. If ω^2 and ω'^2 are the roots of 8.15, it may be written as the real part of

$$q_1 = A_1 e^{i\omega t} + A_1' e^{i\omega' t}, q_2 = A_2 e^{i\omega t} + A_2' e^{i\omega' t}. \tag{8.17}$$

in which the ratios $\frac{A_1}{A_2}$ and $\frac{A_1'}{A_2'}$ are fixed by 8.13.

In the case of the double pendulum, the equations 8.13 are

$$\begin{pmatrix} \frac{M+m}{ML}g + \frac{mg}{Ml} & -\frac{mg}{Ml} \\ -\frac{g}{l} & \frac{g}{l} \end{pmatrix} \begin{pmatrix} A_x \\ A_y \end{pmatrix} = \omega^2 \begin{pmatrix} A_x \\ A_y \end{pmatrix}. \tag{8.18}$$

The characteristic equation 8.15 simplifies to

$$\omega^4 - \frac{M+m}{M}(\frac{g}{L} + \frac{g}{l})\omega^2 + \frac{M+m}{M}\frac{g^2}{Ll} = 0. \tag{8.19}$$

The roots of this equation determine the frequencies of the two normal modes.

It is interesting to examine certain special cases. First, let us suppose that the upper pendulum is very heavy. Then, provided that l is not too close to L, the two roots, with the corresponding ratios determined by 8.18 are approximately,

$$\omega^2 \approx \frac{g}{l}, \frac{A_x}{A_y} \approx \frac{m}{M}\frac{L}{l-L}, \tag{8.20}$$

and

$$\omega^2 \approx \frac{g}{l}, \frac{A_x}{A_y} \approx \frac{L-l}{L}. \tag{8.21}$$

In the first mode, the upper pendulum is practically stationary, while the lower one is swinging with its natural frequency. In the second mode, whose frequency is that of the upper pendulum, the amplitudes are of comparable magnitude.

At the other extreme, if $M \ll m$, the two normal modes are

$$\omega^2 \approx \frac{g}{L+l}, \frac{A_x}{A_y} \approx \frac{L}{L+l}. \tag{8.22}$$

and

$$\omega^2 \approx \frac{m}{M}(\frac{g}{L}+\frac{g}{l}), \frac{A_x}{A_y} \approx -\frac{m}{M}\frac{L+l}{L}. \tag{8.23}$$

In the first mode, the pendulums swing like a single rigid pendulum of length $L+l$. In the second, the lower bob remains almost stationary, while the upper one executes a very rapid oscillation.]

The normal modes of a system with n degrees of freedom may be found by a very similar method. The condition for consistency of the simultaneous equations 8.12 is that the determinant of the coefficients should vanish. For example, for $n = 3$, we require

$$\begin{vmatrix} k_{11} - \omega^2 & k_{12} & k_{13} \\ k_{21} & k_{22} - \omega^2 & k_{23} \\ k_{31} & k_{32} & k_{33} - \omega^2 \end{vmatrix} = 0. \tag{8.24}$$

This is a cubic equation for ω^2. It can be proved that its three roots are all real. As before, the condition for stability is that all three roots should be positive. The roots then determine the frequencies of the three normal modes.

For each normal mode, the ratios of the amplitudes are fixed by the equation 8.12. As in the case $n = 2$, they aer all real, so that in a normal mode the coordinates oscillate in phase. Each normal mode solution involves just two arbitrary constants, and the general solution is a superposition of all the normal modes.

8.3 Coupled oscillation

One often encounters examples of physical system which may be described as tow harmonic oscillators, which are approximately independent, but with some kind of relatively weak coupling between the two.

If the coordinate q of a harmonic oscillator is normalized so that $T = \frac{1}{2}\dot{q}^2$, then $V = \frac{1}{2}\omega^2 q^2$, where ω is the angular frequency. Hence for a pair of uncoupled oscillators, the coefficients in 8.3 are $k_{11} = \omega_1^2$, $k_{12} = 0$ and $k_{22} = \omega_2^2$. When the oscillators are weakly coupled these equalities will still be approximately true, so that in particular k_{12} is small in comparison to k_{11} and k_{22}. Thus from 8.15 it is clear that the characteristic frequencies of the coupled system are given by $\omega^2 \approx k_{11}$ and $\omega^2 \approx k_{22}$. As one might expect, they are close to the frequencies of the uncoupled oscillators.

Then from 8.13 we see that in the first normal mode the ratio $\frac{A_1}{A_2}$ is approximately

$\frac{k_{12}}{(k_{11}-k_{22})}$. Thus unless the frequencies of the two oscillators are nearly equal, the

normal modes differ very little from those of the uncoupled system, and the coupling

is not of great importance. The interesting case, in which even a weak coupling can

be important, is that in which the frequencies are equal, or nearly so.

As a specific example of this case, we consider a pair of pendulums, each of mass

m and length l, coupled by a weak spring. We shall use the displacements x and

y as generalized coordinates. Then in the absence of coupling, the potential energy

is approximately $\frac{1}{2}m\omega_0^2(x^2 + y^2)$, where $\omega_0^2 = \frac{g}{l}$ gives the free oscillation frequency.

The potential energy of the spring has the form $\frac{1}{2}k(x^2 - y^2)$. It will be convenient to

introduce another frequency ω_s defined by $\omega_s = \frac{k}{m}$. In fixed and the other attached

to a mass m. Thus we take

$$V = \frac{1}{2}m(\omega_0^2 + \omega_s^2)(x^2 + y^2) - m\omega_s^2 xy. \tag{8.25}$$

The normal mode equations8.13 now read

$$\begin{pmatrix} \omega_0^2 + \omega_s^2 & -\omega_s^2 \\ -\omega_s^2 & \omega_0^2 + \omega_s^2 \end{pmatrix} \begin{pmatrix} A_x \\ A_y \end{pmatrix} = \omega^2 \begin{pmatrix} A_x \\ A_y \end{pmatrix}. \tag{8.26}$$

The two solutions of the characteristic equation are easily seen to be $\omega^2 = \omega_0^2$

and $\omega^2 = \omega_0^2 + 2\omega_s^2$.

In the first normal mode, the two pendulums oscillate together with equal am-

plitude. Since the spring is neither expanded nor compressed in this motion, it is

not surprising that the frequency is just that of the uncoupled pendulums, In the second normal mode, which has a higher frequency, the pendulum swing in opposite directions, alternately expanding and compressing the spring.

The general solution is a superposition of these two normal modes, and may be written as the real part of

$$x = Ae^{i\omega_0 t} + A'e^{i\omega' t}, y = Ae^{i\omega_0 t} - A'e^{i\omega' t}. \tag{8.27}$$

where $\omega'^2 = \omega_0^2 + 2\omega_s^2$. The constants A and A' may be determined by the initial conditions. For example, if the system is released from rest with one pendulum displaced a distance a from its equilibrium position, so that at $t = 0$ we have $x = a, y = 0, \dot{x} = \dot{y} = 0$, then we easily find that the solution is

$$x = \frac{1}{2}a \cos \omega_0 t + \frac{1}{2}a \cos \omega' t, y = \frac{1}{2}a \cos \omega_0 t - \frac{1}{2}a \cos \omega' t. \tag{8.28}$$

Using standard trigonometrical identities, it may be written

$$x = a \cos \omega_- t \cos \omega_+ t, y = a \sin \omega_- t \sin \omega_+ t. \tag{8.29}$$

where $\omega_\pm = \frac{1}{2}(\omega' \pm \omega_0)$.

Since the spring is weak, ω' is only slightly greater than ω_0 , and therefore $\omega_- \ll \omega_+$. Thus we may describe the motion as follows. the first pendulum swings with angular frequency ω_+, and gradually decreasing amplitude $a \cos \omega_- t$. Meanwhile, the second pendulum starts to swing with the same frequency, but 90 degree out of phase, and with gradually increasing amplitude $a \sin \omega_- t$. After a time $\frac{\pi}{2\omega_-}$, the first pendulum has come momentarily to rest, and the second is oscillating with amplitude a. The whole process is then repeated indefinitely.

This behavior should be contrasted with that of a pair of coupled oscillators of very different frequencies. In such a case, if one is started oscillating, one of the two normal modes will have a much larger amplitude than the other. Thus only a very small oscillation will be set up in the second oscillator, and the amplitude of the first will be pratically constant.

8.3.1 Normal coordinates

The two normal modes of this system are completely independent. We can make this fact explicit by introducing a new pair of coordinates in place of x and y. Let us set

$$q_1 = \frac{m^{1/2}}{\sqrt{2}}(x+y), q_2 = \frac{m^{1/2}}{\sqrt{2}}(x-y). \tag{8.30}$$

In terms of these coordinates , the solution 8.27 is

$$q_1 = A_1 e^{i\omega_0 t}, A_1 = (2m)^{1/2}A, \tag{8.31}$$

$$q_2 = A_2 e^{i\omega' t}, A_2 = (2m)^{1/2}A'. \tag{8.32}$$

Thus in each normal mode one coordinate only is oscillating. Coordinates with this property are called normal coordinates.

The independence of the two normal coordinates may also be seen by examining the Lagrangian function. In terms of q_1 and q_2, the kinetic energy is $T = \frac{1}{2}\dot{q}_1^2 + \frac{1}{2}\dot{q}_2^2$, while the potential energy function 8.25 is $\frac{1}{2}\omega_0^2(q_1^2 + q_2^2) + \omega_s^2 q_2^2$. Thus

$$L = \frac{1}{2}(\dot{q}_1^2 - \omega_0^2 q_1^2) + \frac{1}{2}(\dot{q}_2^2 - \omega'^2 q_2^2). \tag{8.33}$$

In effect, we have reduced the Lagrangian to that of for a pair of uncoupled oscillators with angular frequencies ω_0 and ω'.

The normal coordinates are very useful in studying the effect on the system of a prescribed external force. For example, suppose that the first pendulum is subjected to a periodic force $F(t) = F_1 e^{i\omega_1 t}$. To find the equation of motion in the presence of this force, we must evaluate the work done in a small displacement. This is

$$F(t)\delta x = \frac{F(t)}{(2m)^{1/2}}(\delta q_1 + \delta q_2). \tag{8.34}$$

Thus the equation of motion are

$$\ddot{q}_1 = -\omega_0^2 q_1 + \frac{F_1 e^{i\omega_1 t}}{(2m)^{1/2}}, \tag{8.35}$$

$$\ddot{q}_2 = -\omega'^2 q_2 + \frac{F_1 e^{i\omega_1 t}}{(2m)^{1/2}}, \tag{8.36}$$

These independent oscillator equations may be solved. In particular, the amplitudes of the forced oscillators are given by

$$A_1 = \frac{\frac{F_1}{(2m)^{1/2}}}{\omega_0^2 - \omega_1^2}, A_2 = \frac{\frac{F_1}{(2m)^{1/2}}}{\omega'^2 - \omega_1^2}, \tag{8.37}$$

Note that if the forcing frequency is very close to ω_0, the first normal mode will predominate, and the pendulums will swing in the same direction; while if it is close to ω' the second mode will be more important.

8.4   Oscillations of particles on a string

Consider a light string of length $(n+1)l$, stretched to a tension F, with n equal masses m spaced along it at regular intervals l. We shall consider transverse oscillations of the particles, and use as our generalized coordinates the displacements $y_1, y_1, \cdots, y_n$. Since the kinetic energy is

$$T = \frac{1}{2}m(\dot{y}_1^2 + \dot{y}_2^2 + \cdots + \dot{y}_n^2). \tag{8.38}$$

these coordinates are orthogonal.

Next, we must calculate the potential energy. Let us consider the length of string between the jth and $(j+1)$th particles. In equalibrium, its length is l, but when the particles are displaced it is

$$l + \delta l = [l^2 + (y_{j+1} - y_j)^2]^{1/2} \approx [1 + \frac{(y_{j+1} - y_j)^2}{2l^2}. \tag{8.39}$$

This applies also to the sections of the string at each end if we set $y_0 = y_{n+1} = 0$. The work done against the tension in increasing the length of the string by this amount is $F\delta l$. Thus, adding the contributions from each piece of the string, we find that the potential energy is

$$V = \frac{F}{2l}[y_1^2 + (y_2 - y_1)^2 + \cdots + (y_n - y_{n-1})^2 + y_n^2]. \tag{8.40}$$

It is worth noting that the potential energy of a continuous string may be obtained as a limiting case, as $n \to \infty$ and $l \to 0$. For small l, $\frac{(y_{j+1} - y_j)^2}{l^2}$ is approximately y'^2.

From 8.38 and 8.40, we find that Lagrange's equations are

$$\ddot{y}_1 = \frac{F}{ml}(-2y_1 + y_2), \ddot{y}_2 = \frac{F}{ml}(-y_1 - 2y_2 + y_3), \cdots, \ddot{y}_n = \frac{F}{ml}(y_{n-1} - 2y_n). \quad (8.41)$$

It will be convenient to write $\omega_0^2 = \frac{F}{ml}$. Then substituting the normal mode solution $y_j = A_j e^{i\omega t}$, we obtain the equations

$$\begin{pmatrix} 2\omega_0^2 & -\omega_0^2 & 0 & \cdots & 0 & 0 \\ -\omega_0^2 & 2\omega_0^2 & -\omega_0^2 & \cdots & 0 & 0 \\ \cdots & \cdots & \cdots & \cdots & \cdots & \cdots \\ 0 & 0 & 0 & \cdots & -\omega_0^2 & 2\omega_0^2 \end{pmatrix} \begin{pmatrix} A_1 \\ A_2 \\ \cdots \\ A_n \end{pmatrix} = \omega^2 \begin{pmatrix} A_1 \\ A_2 \\ \cdots \\ A_n \end{pmatrix}. \quad (8.42)$$

For $n = 1$, there is one normal mode, with $\omega^2 = 2\omega_0^2$. For $n = 2$, the characteristic equation is

$$(2\omega_0^2 - \omega^2)^2 - \omega_0^4 = 0, \quad (8.43)$$

and we obtain two normal modes with

$$\omega^2 = \omega_0^2, \frac{A_1}{A_2} = 1. \quad (8.44)$$

and

$$\omega^2 = 3\omega_0^2, \frac{A_1}{A_2} = -1. \quad (8.45)$$

Now let us examine the case $n = 3$. The characteristic equation is

$$\begin{vmatrix} 2\omega_0^2 - \omega^2 & -\omega^2 & 0 \\ -\omega_0^2 & 2\omega_0^2 - \omega^2 & \omega_0^2 \\ 0 & \omega_0^2 & 2\omega_0^2 - \omega^2 \end{vmatrix} = 0. \qquad (8.46)$$

Expanding this determinant by the usual rules, we obtain a cubic equation for ω^2,

$$(2\omega_0^2)^3 - 2\omega_0^4(2\omega_0^2 - \omega^2) = 0. \qquad (8.47)$$

The roots of this equation are $2\omega_0^2$ and $(2 \pm \sqrt{2})\omega_j^2$. Hence we obtain three normal modes

$$\omega^2 = (2 - \sqrt{2})\omega_0^2, A_1 : A_2 : A_3 = 1 : \sqrt{2} : 1, \omega^2 = 2\omega_0^2, \qquad (8.48)$$

$$A_1 : A_2 : A_3 = 1 : 0 : -1, \omega^2 = (2 + \sqrt{2})\omega_0^2, A_1 : A_2 : A_3 = 1 : -\sqrt{2} : 1, \qquad (8.49)$$

Higher values of n may be treated similarly. For $n = 4$, the characteristic equation is given by the vanishing of a 4×4 determinant, which may be expanded by similar rules to yield

$$(2\omega_0^2 - \omega^2)^4 - 3\omega_0^4(2\omega_0^2 - \omega^2)^2 + \omega_0^8 = 0. \qquad (8.50)$$

The roots of this equation are given by

$$(2\omega_0^2 - \omega^2)^2 = \frac{3 \pm \sqrt{5}}{2}\omega_0^4 = (\frac{\sqrt{5} \pm 1}{2}\omega_0^2)^2. \qquad (8.51)$$

Thus we obtain four normal modes:

$$\omega^2 = 0.38\omega_0^2, A_1 : A_2 : A_3 : A_4 = 1 : 1.62 : 1.62 : 1; \omega^2 = 1.38\omega_0^2, A_1 : A_2 : A_3 : A_4 = 1.62 : 1 : -$$

$$(8.52)$$

For every value of n , the slowest mode is the one in which all the masses are oscillating in the same direction, while the fastest is one in which alternate masses oscillate in opposite directions. For large values of n, the normal modes approach those of a continuous stretched string.

8.5   Normal modes of a stretched string

The equation of motion of stretched string is

$$\ddot{y} = c^2 y'', c^2 = \frac{F}{\mu} \qquad (8.53)$$

and look for the normal mode solutions of the form

$$y(x, t) = A(x)e^{i\omega t}. \qquad (8.54)$$

Substituting in 8.53, we obtain

$$A''(x) + k^2 A(x) = 0, \qquad (8.55)$$

,where $k = \frac{\omega}{c}$.

Thus, in place of a set of simulataneous equations for the amplitudes A_j, we obtain a differential equation for the amplitude function $A(x)$.

The general solution of this equation is

$$A(x) = a \cos kx + b \sin kx. \tag{8.56}$$

However, because the ends of the string are fixed, we must impose the boundary conditions $A(0) = A(l) = 0$. Thus $a = 0$, and $\sin kl = 0$. The possible values of k are

$$k = \frac{n\pi}{l}, n = 1, 2, 3, \cdots. \tag{8.57}$$

Each of these values of k corresponds to a normal mode of the string. The corresponding angular frequencies are

$$\omega = \frac{n\pi c}{l}, n = 1, 2, 3, \cdots. \tag{8.58}$$

They are all multiples of the fundamental frequency $\frac{\pi c}{l} = \pi(\frac{F}{Ml})^{1/2}$ where M is the total mass of the string.

The solution of the nth normal mode can be written as the real part of

$$y(x, t) = A_n e^{\frac{in\pi ct}{l}} \sin \frac{n\pi x}{l}, \tag{8.59}$$

, where A_n in an arbitrary complex constant. It represents a standing wave of wavelength $\frac{2l}{n}$, with $n - 1$ nodes, or points at which $y = 0$. The general solution for the stretched string is a superposition of all the normal modes 8.59.

Chapter 9

Hamiltonian and Lagrangian

9.1 Generalized coordinates

Let us consider a rigid body composed of a large number N of particles. The positions of all the particles may be specified by 3N coordinates. However these 3n coordinates cannot all vary independently, but are subject to constraints -the rigidly conditions, In fact, the position of every particle may be fixed by specifying the values of just six quantities- for instance, the three coordinates X, Y, Z of the center of mass. and the three Euler angles ϕ, θ, ψ which determine the orientation, These six constitute a set of generalized coordinates for the rigid body.

These coordinates may be subject to further constraints, which may be of two kinds. First, we might fix the position of one point of the body. Such constraints are represented by algebraic conditions on the coordinates, which may be used to eliminate some of the coordinates. In this particular case, the three Euler angles alone suffice to fix the position of every particle.

The second type of constraint is represented by conditions on the velocities rather than the coordinates. For example, we might constrain the center of mass to move with uniform velocity, or to move round a circle with uniform angular velocity. Then in place of algebraic equations. we have differential equations. In simple cases, these equations can be solved to find some of the coordinates as explicit functions of time. Then the position of every particle will be determined by the values of the remaining generalized coordinates and the time t.

In general, we say that $q_1, q_2, \cdots, q_n$ is a set of generalized coordinates for a given system if the position of every particle in the system is a function of these variables, and perhaps also explicitly of time,

$$\mathbf{r}_i = \mathbf{r}_i(q_1, q_2, \cdots, q_n, t). \tag{9.1}$$

The number of coordinates which can vary independently is called the number of degrees of freedom of the system. If it is possible to solve the constraint equations, and eliminate some of the coordinates, leaving a set equal in number to the number of degrees of freedom, the system is called "holonomic". If this elimination introduces explicit functions of time, the system is said to be forced; on the other hand, if all the constraints are purely algebraic, so that t does not appear explicitly in 9.1 , the system is natural.

There exist non-holomic system, for which the constraint equations cannot be solved to eliminate some of the coordinates. Consider a sphere rolling on a rough plane. Its position may be specified by five generalized coordinates, X, Y, ϕ, θ, ψ (Z is a constant, and may be omitted). However, the sphere can only roll in two directions, and the number of degrees of freedom is two. The constraint equations serve to determine the angular velocity in terms of the velocity of center of mass. Using the fact that the instantaneous axis of rotation must be a horizontal axis through the point of contact, it is not hard to show that $\omega = a\mathbf{k} \times \dot{\mathbf{R}}$. However these equations cannot be integrated to find the orientation in terms of the position of the center of mass; because we could roll the sphere round a circle so that it returns to its starting point but with a different orientation.

The distinction between a natural and a forced system can be expressed in another way, which will be useful later. Differentiating 9.1 with respect to the time, we find that the velocity of the ith particle is a linear function of $\dot{q}_1, \dot{q}_2, \cdots, \dot{q}_n$, though in general it depends in a more complicated way on the coordinates $q_1, \cdots, q_n$ themselves:

$$\dot{\mathbf{r}} = \Sigma_{\alpha=1}^{n} \frac{\partial \mathbf{r}_i}{\partial q_\alpha} \dot{q}_\alpha + \frac{\partial \mathbf{r}_i}{\partial t} \tag{9.2}$$

The last term arises from the explicit dependence on t, and is absent for a natural system. When we substitute in $T = \Sigma \frac{1}{2} m \dot{\mathbf{r}}^2$, we obtain a quadratic function of the time derivatives $\dot{q}_1, \dot{q}_2, \cdots, \dot{q}_n$. For a natural system, it is a homogeneous quadratic function; but for a forced system there are also linear and constant terms.

For example, the kinetic energy of a symmetric rigid body is

$$T = \frac{1}{2}M(\dot{X}^2 + \dot{Y}^2 + \dot{Z}^2) + \frac{1}{2}I_1^*(\dot{\phi}^2 \sin^2\theta + \dot{\theta}^2) + \frac{1}{2}I_3^*(\dot{\psi}\cos\theta + \dot{\psi}^2). \tag{9.3}$$

If we impose further algebraic constraints, such as $X = 0$, the corresponding terms drop out, and we are still left with a homogeneous quadratic function of the remaining time derivatives. On the other hand, if we impose differential constraints, such as $\dot{X} = v$, or $\dot{\phi} = \omega$ we obtain a function with constant or linear terms.

9.2 Calculus of Variations

It will be helpful to begin by discussing a very simple example. Let us ask the question; what is the shortest path between two given points in a plane? Or course we know the answer already, but the method we shall use to derive it can also be applied to less trivial examples - for instance, to find the shortest path between two points on a curved surface.

Suppose the two points are (x_0, y_0) and (x_1, y_1). Any curve joining them is represented by an equation

$$y = y(x) \tag{9.4}$$

such that the function $y(x)$ satisfies the boundary conditions

$$y(x_0) = y_0, y(x_1) = y_1. \tag{9.5}$$

Consider two neighbouring points on this curve. The distance dl between them is given by

$$dl = (dx^2 + dy^2)^{1/2} = (1 + y'^2)^{1/2} dx \tag{9.6}$$

where $y' = \frac{dy}{dx}$. Thus the total length of the curve is

$$l = \int_{x_0}^{x_1} (1 + y'^2)^{1/2} dx \tag{9.7}$$

Then the problem is to find that function $y(x)$, subject to the conditions 9.5, which will make this integral a minimum.

This problem differs from the usual kind of minimum-value problem in that what we have to vary is not a single variable or set of variables, but a function $y(x)$. However, we can still apply the same criterion: when the integral has a minimum value, it must be unchanged to first order by making a small variation in the function $y(x)$.

More generally, we may be interested in finding the stationary values of an integral of the form

$$I = \int_{x_0}^{x_1} f(y, y') dx, \tag{9.8}$$

where $f(y, y')$ is a specified function of y and its first derivative. We shall solve this general problem, and then apply the result to the integral 9.7. Consider a small variation $\delta y(x)$ in the function $y(x)$, subject to the condition that the values of y at the end-points are unchanged :

$$\delta y(x_0) = 0, \delta y(x_1) = 0. \tag{9.9}$$

To first order, the variation in $f(y, y')$ is

$$\delta f = \frac{\partial f}{\partial y} + \frac{\partial f}{\partial y'} \delta y', \tag{9.10}$$

where

$$\delta y' = \frac{d}{dx}\delta y. \tag{9.11}$$

Thus the variation of the integral I is

$$\delta I = \int_{x_0}^{x_1}[\frac{\partial f}{\partial y}\delta y + \frac{\partial f}{\partial y'}\frac{d}{dx}\delta y]dx. \tag{9.12}$$

In the second term, we may integrate by parts. The integrated term, namely

$$[\frac{\partial f}{\partial y'}\delta y]_{x_0}^{x_1} \tag{9.13}$$

vanished at the limits because of the conditions. Hence we obtain

$$\delta I = \int_{x_0}^{x_1}[\frac{\partial f}{\partial y} - \frac{d}{dx}(\frac{\partial f}{\partial y'})]\delta y(x)dx. \tag{9.14}$$

In order that I should be stationary, this variation δI must vanish for an arbitrary small variation $\delta y(x)$. This is only possible if the integrand vanished identically. Thus we require

$$\frac{\partial f}{\partial y} - \frac{d}{dx}(\frac{\partial f}{\partial y'}) = 0. \tag{9.15}$$

This is known as the Euler-Lagrange equation. It is general a second order differential equation for the function $y(x)$, whose solution contains two arbitrary constants that may be determined from the known values of y at x_0 and x_1.

Now we can solve the problem. In that case, comparing 9.7 and ??, we have to choose

$$f = (1 + y'^2)^{1/2} \tag{9.16}$$

and therefore

$$\frac{\partial f}{\partial y} = 0, \frac{\partial f}{\partial y'} = \frac{y'}{(1 + y'^2)^{1/2}}. \tag{9.17}$$

Thus the Euler-Lagrange equation reads

$$\frac{d}{dx}\left[\frac{y'}{(1 + y'^2)^{1/2}}\right] = 0. \tag{9.18}$$

This equation states that the expression inside the bracket is a constant, and hence that y' is a constant. Its solutions are therefore the straight lines.

$$y = ax + b \tag{9.19}$$

Thus we have proved that the shortest path between two points is a straight line.

So far, we have used x as the independent variable, but in the applications we consider later we shall be concerned instead with functions of the time t. It is easy to generalize the discussion to the case of function f of n variables $q_2, \cdots, q_n$, and their time derivatives $\dot{q}_1, \dot{q}_2, \cdots, \dot{q}_n$. In order that the integral

$$I = \int_{t_0}^{t_1} f(q_1, \cdots, q_n, \dot{q}_1, \cdots, \dot{q}_n) dt \tag{9.20}$$

be stationary, it must be unchanged to first order by a variation in any one of the functions $q_i(t)$, subject to the conditions $\delta q_i(t_0) = \delta q_i(t_1) = 0$. Thus we require the n Euler-Lagrange equations

$$\frac{\partial f}{\partial q_i} - \frac{d}{dt}\left(\frac{\partial f}{\partial \dot{q}_i}\right) = 0, \tag{9.21}$$

These n second-order differential equation determine the n functions $q_i(t)$ to within $2n$ arbitrary constants of integration.

9.3   Hamilton's principle

Consider the one dimensional motion of a particle with coordinate x in a potential $V(x)$. Call the particle's kinetic energy

$$T(\dot{x}) = \frac{m}{2}\dot{x}^2. \tag{9.22}$$

Now we would like to find a particular trajectory $x(t)$ are such that the particle moves from x_1 at time t_1 to x_2 at time t_2,

$$x(t_1) = x_1, x(t_2) = x_2. \tag{9.23}$$

Hamilton's principle is the statement that if you compute the quantity

$$S[x] \equiv \int_{t_1}^{t_2} dt[T(\dot{x}(t)) - V(x(t))]. \tag{9.24}$$

depending on a function $x(t)$ satisfying then the variation of $S[x]$ with respect to the function $x(t)$ vanishes for the actual trajectory. The hard part of Hamilton's principle will not be proving this statement, but figuring out exactly what it means to talk about the variation of something with respect to a function.

The first thing to notice is that $T - V$ is an ordinary function of two variables, x and $\dot{x}$, but that $S[x]$ is actually a function of a function.

Thus $S[x]$ is a function whose argument is itself a function. I have put the x in square brackets to remind you that in this case x is a function rather than a number. The value of $S[x]$, on the other hand, is just a number. It doesn't depend on t. the variable t is a dummy variable.

Now Hamilton's principle is a statement about the variation of $S[x]$ as we let $x(t)$ vary over all possible functions. We will find that when $x(t)$ is very near to a solution to Newton's second law, then $S[x]$ changes slowly as a function of whatever parameter you use to specify the function $x(t)$. When $x(t)$ is a solution, $S[x]$ is not varying at all.

9.4 Functions of functions

A functional like $S[x]$ is just a function of an infinite number of variables where the variables are the values of $x(t)$ at all possible values of t. The really peculiar new thing about a functional is that the variables are labled by a continuous parameter, rather than having different names like x, y and z, or different discrete indices, like a_1, a_2,etc. This is why functionals look so different and why we have to invent some new notation to deal with them.

A mathematical example of a functional is the length of a path described by a curve, $y = f(x)$. the path length from x_1 to x_2 is

$$P[f] = \int_{x_1}^{x_2} \sqrt{dx^2 + dy^2} = \int_{x_1}^{x_2} dx\sqrt{1 + (\frac{dy}{dx})^2} = int_{x_1}^{x_2} dx\sqrt{1 + f'(x)^2}. \qquad (9.25)$$

The path length depends on the function f that defines the shape of the curve.

9.5   The Lagrange's Equation

Hamilton's principle really captures more of what is going on in the world than $F = ma$. It is worth going over the consequences of Hamilton's principle again in a more general language. In general, the combination $T - V$ is called the Lagrangian, L. $S[x]$ is called the action,

$$S[x] = \int_{t_1}^{t_2} dt L(x(t), \dot{x}(t)). \qquad (9.26)$$

Here, x might actually have indices that allows it to represent more than one particle or dimension or both. We won't write them explicitly in this formal derivation. Now we want the functional derivative of $S[x]$ with respect to $x(t)$ to vanish. In general, the functional derivative is

$$\frac{\delta S}{\delta x(t)}[x] = \frac{\partial}{\partial x(t)}L(x(t), \dot{x}(t)) - \frac{d}{dt}\frac{\partial}{\partial \dot{x}(t)}L(x(t), \dot{x}(t)). \qquad (9.27)$$

The first term arises in the obvious way from the Taylor expansion of the $x(t)$ dependence.

The Hamilton's principle implies that the solution for the motion satisfies

$$\frac{\partial}{\partial x(t)}L(x(t), \dot{x}(t)) - \frac{d}{dt}\frac{\partial}{\partial \dot{x}(t)}L(x(t), \dot{x}(t)). \qquad (9.28)$$

It x has several components, the above equation must be true for each component separately.

This equation is called the Euler-Lagrange equation. We have seen that in simple situation this equation is just a trivial rewriting of $F = ma$. However, there are several important advantages to thinking about mechanics in terms of Hamilton's principle, and deriving the Euler-Lagrange equation, rather than using Newton's second law directly.

1. It is easier to write down the kinetic and potential energies than to understand all of the forces. Thus Hamilton's principle can be a labor saving device.

2. In general, there may be many different ways of representing the configuration of a physical system, and thus many ways of choosing the variables that describe the system. Hamilton's principle makes it obvious that the particular choice of variable doesn't matter, because $S[x]$, whose vanishing variation determines the trajectories, is just a single number that doesn't depend on how we choose to describe the system. This will be especially useful when we work in spherical or cylindrical coordinate, which are more appropriate to some problems.

3. More generally, Hamilton's principle make it very easy to understand the consequences of symmetry. We will use it to see that there is a deep connection between symmetries and conservation laws like conservation of momentum or angular momentum.

4. Sometimes, we don't know or care about all the forces in a mechanical system. This often happens when there is a constraint. There are many examples, like a bead sliding on a frictionless wire, or a particle moving on the surface of the earth,

where the system is constrained by some forces that we don't understand in detail. We can use Hamilton's principle to solve such problems without even computing the constraining forces by simply choosing coordinates that automatically incorporate the constraints.

To solve many actual problems we begin by constructing the Lagrangian L.

$$L(x(t), \dot{x}(t), t) = T(x(t), \dot{x}(t), t) - U(x(t), \dot{x}(t), t). \tag{9.29}$$

Where T and U are the kinetic and potential energies, as functions of the variable x that describes the configuration of the system, and its time derivative, $\dot{x}$. This time we have included the possibility that these functions also depend explicitly on t. By explicit time dependence, I mean dependence beyond that coming from the time dependence of $x(t)$.

From the Lagrangian, we construct the action $S[x]$ by integrating in time from an initial time t_1 to a final time t_2.

$$S[x] = \int_{t_1}^{t_2} dt L(x(t), \dot{x}(t), t). \tag{9.30}$$

Then Hamilton's principle is the statement that the actual trajectory that describes the evolution of the system from $x(t_1) = x_1$ to $x(t_2) = x_2$ is the trajectory around which the variation of $S[x]$ vanishes. What I showed you last time is that we could turn this condition into a differential equation for the trajectory by setting to zero the functional derivative

$$\frac{\delta S[x]}{\delta x(t)} = \frac{\partial}{\partial x(t)} L(x(t), \dot{x}(t), t) - \frac{d}{dt} \frac{\partial}{\partial \dot{x}(t)} L(x(t), \dot{x}(t), t). \tag{9.31}$$

Thus Hamilton's principle implies that the solution for the motion satisfies

$$\frac{\partial}{\partial x(t)} L(x(t), \dot{x}(t), t) - \frac{d}{dt} \frac{\partial}{\partial \dot{x}(t)} L(x(t), \dot{x}(t), t) = 0. \tag{9.32}$$

This is the Euler-Lagrangian equation. Hamilton's principle tell us that the classical trajectory satisfies the above relation. But we are usually interested not in checking that a given solution satisfies the Euler-Lagrange equation, but rather in using the Euler- Lagrange equation to find the solution. Then the above equation is just a differential equation for $x(t)$, it is convenient to drop the explicit time dependence of x and write it as

$$\frac{\partial}{\partial x} L(x, \dot{x}, t) - \frac{d}{dt} \frac{\partial}{\partial \dot{x}} L(x, \dot{x}, t) = 0. \tag{9.33}$$

Solving the differential equation and imposing the initial condition then gives us $x(t)$.

9.6 Generalized Force and Momentum

For a particle of mass m moving in a potential $V(x)$, the Euler-Lagrange equation of motion can be written as

$$\frac{dp}{dt} = \frac{d}{dt}(m\dot{x}) = \frac{d}{dt} \frac{\partial L}{\partial \dot{x}} = \frac{\partial L}{\partial x} = -V'(x) = F(x). \tag{9.34}$$

The rate of change of the momentum is equal to the force. In the general situation, this suggests that we might regard the Euler-Lagrange equations,

$$\frac{d}{dt}\frac{\partial L}{\partial \dot{q}_j} = \frac{\partial L}{\partial q_j} \tag{9.35}$$

as a generalization of this, we call $\frac{\partial L}{\partial \dot{q}_j}$ the "generalized momentum" corresponding to the coordinate q_j and $\frac{\partial L}{\partial q_j}$ the "generalized force" corresponding to the coordinate q_j. Then the Lagrange equation says that the rate of change of the generalized momentum equals the corresponding generalized force.

A particular interesting case occurs when the Lagrangian does not depend at all on some coordinate q_j. In that case, equation XX implies that the generalized momentum corresponding to q_j is constant. This statement becomes even more interesting when you realize that we have great freedom to choose the coordinates any way we want to. Thus if there is any coordinate system in which the Lagrangian does not depend on some coordinate, then there is a conservation law- the corresponding generalized momentum is conserved.

9.7   Charged particle in an electromagnetic field

We consider one of the most important examples of a non-conservative force. Let's assume that a particle of charge q is moving in an electric field $\mathbf{E}$ and magnetic field $\mathbf{B}$. The force on the particle is then

$$\mathbf{F} = q(\mathbf{E} + \mathbf{v} \times \mathbf{B}), \tag{9.36}$$

or, in terms of components,

$$F_x = qE_x + q(\dot{y}B_z - \dot{z}B_y), \tag{9.37}$$

with two similar equations obtained by cyclic permutation of x,y,z.

Now we wish to show that this force may be written with a suitably chosen function V. To do this, we have to make use of a standard result of electromagnetic theory, according to which it is possible to find a scalar potential ϕ, and a vector potential $\mathbf{A}$, functions of $\mathbf{r}$ and t such that

$$\mathbf{E} = -\nabla\phi - \frac{\partial \mathbf{A}}{\partial t}, \tag{9.38}$$

$$\mathbf{B} = \nabla \times \mathbf{A}. \tag{9.39}$$

For time-independent fields, the scalar potential ϕ is simply the electromagnetic potential.

Now let us consider the function

$$V = q\phi(\mathbf{r}, t) - q\dot{\mathbf{r}} \cdot \mathbf{A}(\mathbf{r}, t) = a\phi - q(\dot{x}A_x + \dot{y}A_y + \dot{z}A_z) \tag{9.40}$$

Clearly,

$$-\frac{\partial v}{\partial x} = -q\frac{\partial \phi}{\partial x} + a(\dot{x}\frac{\partial A_x}{\partial x} + \dot{y}\frac{\partial A_y}{\partial x} + \dot{z}\frac{\partial A_z}{\partial x}). \tag{9.41}$$

Also,

$$\frac{d}{dt}(\frac{\partial V}{\partial \dot{x}}) = -q\frac{dA_x}{dt} = -q(\frac{\partial A_x}{\partial t} + \frac{\partial A_x}{\partial x}\dot{x} + \frac{\partial A_x}{\partial y}\dot{y} + \frac{\partial A_x}{\partial z}\dot{z}) \tag{9.42}$$

Since A_x varies with time both because of its explicit time dependence, and because of its dependence on the particle position $\mathbf{r}$. Hence, adding, we obtain

$$-\frac{\partial V}{\partial x} + \frac{d}{dt}\left(\frac{\partial V}{\partial \dot{x}}\right) = a\left(-\frac{\partial \phi}{\partial x} - \frac{1}{c}\frac{\partial A_x}{\partial t}\right) + q\left[\dot{y}\left(\frac{\partial A_y}{\partial x} - \frac{\partial A_x}{\partial y}\right) + \dot{z}\left(\frac{\partial A_z}{\partial x} - \frac{\partial A_x}{\partial z}\right)\right] \quad (9.43)$$

$$= qE_x + q(\dot{y}B_z - \dot{z}B_y) \quad (9.44)$$

It follows that the equations of motion for a particle in an electromagnetic field may be obtained from the Lagrangian function

$$L = \frac{1}{2}m\dot{\mathbf{r}}^2 + q\dot{\mathbf{r}} \cdot \mathbf{A}(\mathbf{r}, t) - q\phi(\mathbf{r}, t) \quad (9.45)$$

Note the appearance in L of terms linear in the time derivatives. This function cannot be separated into two parts, one quadratic in $\dot{\mathbf{r}}$, and one independent of it. Another consequence of the appearance of these terms is that the generalized momentum p_x is no longer equal to the familiar mechanical momentum $m\dot{x}$. Instead,

$$p_x = \frac{\partial L}{\partial \dot{x}} = m\dot{x} + qA_x \quad (9.46)$$

or, more generally

$$\mathbf{p} = m\dot{\mathbf{r}} + q\mathbf{A} \quad (9.47)$$

We can now obtain the equation of motion in terms of arbitrary coordinates from the Lagrangian function. For example, in terms of cylindrical polars, it reads

$$L = \frac{1}{2}m(\dot{\rho}^2 + \rho^2\dot{\phi}^2 + \dot{z}^2) + q(\dot{\rho}A_\rho + \rho\dot{\phi}^2 A_\phi + \dot{z}A_z) - q\phi. \qquad (9.48)$$

Let us consider the case of a uniform static magnetic field $\mathbf{B}$. In this case, we may take $\phi = 0$, and $\mathbf{A} = \frac{1}{2}\mathbf{B} \times \mathbf{R}$, or if $\mathbf{B}$ is in the z direction

$$A_\rho = 0, A_\phi = \frac{1}{2}B\rho, A_z = 0. \qquad (9.49)$$

Thus the Lagrangian function is

$$L = \frac{1}{2}m(\dot{\rho}^2 + \rho^2\dot{\phi}^2 + \dot{z}^2) + \frac{1}{2}qB\rho^2\dot{\phi}. \qquad (9.50)$$

Therefore Lagrange's equations are

$$m\ddot{\rho} = m\rho\dot{\phi}^2 + qB\rho\dot{\phi}, \qquad (9.51)$$

$$\frac{d}{dt}(m\rho^2\dot{\phi} + \frac{1}{2}qB\rho^2) = 0 \qquad (9.52)$$

$$m\ddot{z} = 0 \qquad (9.53)$$

In particular, let us find the solutions in which ρ is constant. In that case, we learn form the last two equations that $\dot{\phi}$ and $\dot{z}$ are also constants, and form the first equation that either $\dot{\phi} = 0$ or

$$\dot{\phi} = -\frac{qB}{m}. \qquad (9.54)$$

The equation of motion 9.52 is particularly interesting. It shows that although in general the z component of the particle angular momentum J_z is not a constant, there is still a corresponding conservation law for the quantity.

$$p_\phi = m\rho^2\dot{\phi} + \frac{1}{2}qB\rho^2. \tag{9.55}$$

9.8   Pendulum constrained to rotate about an axis

Let us consider the system which consists of a light rigid rod of length l, carrying a mass m at one end, and hinged at the other end to a vertical axis, so that it can swing freely in a vertical plane. We suppose first that a known torque G is applied to this axis.

The system has two degrees of freedom. Its position may be described by the two polar angles θ, ϕ. The Lagrangian function is

$$L = \frac{1}{2}ml^2(\dot{\theta}^2 + \dot{\phi}^2\sin^2\theta) - mgl(1 - \cos\theta). \tag{9.56}$$

Since the work done by the torque G is $\delta W = G\delta\phi$, Lagrange's equations are

$$\frac{d}{dt}\left(\frac{\partial L}{\partial\dot{\theta}}\right) = \frac{\partial L}{\partial\theta}, \frac{d}{dt}\left(\frac{\partial L}{\partial\dot{\phi}}\right) = \frac{\partial L}{\partial\phi} + G, \tag{9.57}$$

or, explicitly

$$ml^2\ddot{\theta} = ml^2\dot{\phi}^2\sin\theta\cos\theta - mgl\sin\theta. \tag{9.58}$$

and

$$\frac{d}{dt}(ml^2\dot{\phi}\sin^2\theta) = G. \tag{9.59}$$

Now let us suppose that the torque G is adjusted to constrain the system to rotate with constant angular velocity ω about the vertical. This imposes the constraint $\dot{\phi} = \omega$, and the system may be regarded as a system with one degree of freedom, described by the coordinate θ. Substituting this constraint in the Lagrangian, we obtain

$$L = \frac{1}{2}ml^2(\dot{theta}^2 + \omega^2\sin^2\theta) - mgl(1 - \cos\theta). \tag{9.60}$$

Note the appearance in the kinetic energy part of L of a term independent of $\dot{\theta}$, which is characteristic of a forced system.

There is a more general way of handling constraints, using the method of Lagrange multipliers. To impose the constraint $\dot{\theta} = \omega$ on our system, we introduce a new variable λ, the Lagrange multiplier, and subtract λ times the constraint form the Lagrangian function. From 9.56 we obtain in this way

$$L = \frac{1}{2}ml^2(\dot{\theta}^2 + \dot{\phi}^2\sin^2\theta) - mgl(1 - \cos\theta) - \lambda(\dot{\phi} - \omega). \tag{9.61}$$

The Euler-Lagrange equation for θ is unchanged. That for ϕ becomes

$$?? \frac{d}{dt}(ml^2\dot{\phi}\sin^2\theta - \lambda) = 0. \tag{9.62}$$

In addition, we have the Euler-Lagange equation for λ, which simply reproduces the constraint

$$0 = \dot{\phi} - \omega \tag{9.63}$$

Comparing **??** with 9.58, we see that $\dot{\lambda}$ is the torque G required to impose the constraint. Equivalently, we could have integrated the constraint, and subtracted $\lambda(\phi - \omega t)$ from L. In that case, λ itself would be the torque. In general, the physical significance of λ may be found by considering a small virtual change in the constraint equation. This method is particularly useful when we want to find the constraining forces or torques.

9.9 Hamilton's equations

The Lagrangian function L is a function of $q_1, q_2, \cdots, q_n$ and $\dot{q}_1, \dot{q}_2, \cdots, \dot{q}_n$. For brevity, we shall indicate this dependence by writing $L(q, \dot{q})$. where q stands for all the generalized coordinates, and $\dot{q}$ for all their time derivatives.

Lagrange's equations may be written in the form

$$\dot{p}_\alpha = \frac{\partial L}{\partial q_\alpha}. \tag{9.64}$$

where the generalized momenta are defined by

$$p_\alpha = \frac{\partial L}{\partial q_\alpha} \tag{9.65}$$

Here, α runs over $1, 2, \cdots, n$.

The instantaneous position and velocity of every part of our system may be specified by the values of the $2n$ variables q and $\dot{q}$. However, we can alternatively solve the equations 9.65 for the $\dot{q}$ in terms of q and p, obtaining,

$$\dot{q}_\alpha = \dot{q}_\alpha(q, p), \tag{9.66}$$

and specify the instantaneous position and velocity by means of the $2n$ variables q and p.

For example, for a particle moving in a plane, and described by polar coordinates, $p_r = mr^2\dot{\theta}$. In this case, the equations 9.66 read

$$\dot{r} = \frac{p_r}{m}, \dot{\theta} = \frac{p_\theta}{mr^2}. \tag{9.67}$$

The instantaneous position and velocity of the particle may be fixed by the values of r, θ, p_r, and p_θ.

Now we define a function of q and p , the Hamiltonian function by

$$H = \sum_{\beta=1}^{n} p_\beta \dot{q}_\beta - L. \tag{9.68}$$

Here the variables $\dot{q}$ are to be regarded as functions of q and p. Written out to indicate the functional dependence, 9.68

$$H(q, p) = \sum_{\beta=1}^{n} p_\beta \dot{q}_\beta(q, p) - L(q, \dot{q}(q, p)). \tag{9.69}$$

Now we compute the derivatives of H. We differentiate first with respect to p_α. One term in this derivatives is the coefficient of p_α in the sum $\Sigma p\dot{q}$, namely $\dot{q}_\alpha$. Other terms arise from the dependence of $\dot{p}$ on p_α. Altogether we obtain,

$$\frac{\partial H}{\partial p_\alpha} = \dot{q}_\alpha + \sum_{\beta=1}^{n} p_\beta \frac{\partial \dot{q}_\beta}{\partial q_\alpha} - \sum_{\beta=1}^{n} \frac{\partial L}{\partial \dot{q}_\beta} \frac{\partial \dot{q}_\beta}{\partial q_\alpha} \tag{9.70}$$

Now by 9.65 the second and third terms cancel. Hence we have

$$\frac{\partial H}{\partial p_\alpha} = \dot{q}_\alpha. \tag{9.71}$$

We examine the derivative with respect to q_α. There are two kinds of terms, the term coming from the explicit dependence of L on q_α, and those from the dependence of $\dot{q}$ on q_α. Thus

$$\frac{\partial H}{\partial p_\alpha} = -\frac{\partial H}{\partial q_\alpha} + \Sigma_{\beta=1}^{n} p_\beta \frac{\partial \dot{q}_\beta}{\partial q_\alpha} - \sum_{\beta=1}^{n} \frac{\partial L}{\partial \dot{q}_\beta} \frac{\partial \dot{q}_\beta}{\partial q_\alpha} \tag{9.72}$$

As before the second and third terms cancel. Using Lagrnage's equations we obtain.

$$\frac{\partial H}{\partial p_\alpha} = -\dot{p}_\alpha. \tag{9.73}$$

The equations 9.71 and 9.73 together constitute Hamilton's equations. Whereas Lagrange's equations are a set of n second-order differential equations, these are a set of $2n$ first-order equations.

Let us consider, a particle moving in a plane under a central conservative force. Then,

$$L = \frac{1}{2}m\dot{r}^2 + \frac{1}{2}mr^2\dot{\theta}^2 - V(r). \tag{9.74}$$

Hence the Hamiltonian function is

$$H = p_r \dot{r} + p_\theta \dot{\theta} - (\frac{1}{2}m\dot{r}^2 + \frac{1}{2}mr^2\dot{\theta}^2 - V(r)). \tag{9.75}$$

or, using 9.67

$$H = \frac{p_r^2}{2m} + \frac{p_\theta^2}{2mr^2} + V(r). \tag{9.76}$$

It may be noticed that his is the expression for the total energy, $T + V$. This is no accident, but a general property of natural systems, as we shall see below.

The first pair of Hamilton's equations are

$$\dot{r} = \frac{\partial H}{\partial p_r} = \frac{p_r}{m}, \dot{\theta} = \frac{\partial H}{\partial p_\theta} = \frac{p_\theta}{mr^2}. \tag{9.77}$$

They simply reproduce the relations 9.67 between velocities and momenta. The second pair are

$$-\dot{p}_r = \frac{\partial H}{\partial r} = -\frac{p_\theta^2}{mr^3} + \frac{dV}{dr}, -\dot{p}_\theta = \frac{\partial H}{\partial \theta} = 0. \tag{9.78}$$

The second of these two equations yields the law of conservation of angular momentum.

$$p_\theta = J = const \tag{9.79}$$

The first gives the radial equation of motion.

$$\dot{p}_r = m\ddot{r} = \frac{J^2}{mr^3} - \frac{dV}{dr}. \tag{9.80}$$

It may be integrated to give the radial energy equation.

9.10   Ignorable coordinates

It sometimes happens that one of the generalized coordinates, say q_α, does not appear in the Hamiltonian function. In that case, the coordinate q_α is said to be ignorable-for reason we shall explain in a moment. For an ignorable coordinate, Hamiltonian's equation

$$-p_\alpha = \frac{\partial H}{\partial q_\alpha} = 0. \tag{9.81}$$

leads immediately to a conservation law for the corresponding generalized momentum,

$$p_\alpha = const. \tag{9.82}$$

For example, for a particle moving in a plane under a central conservative force, H is independent of the angular coordinate θ, and we therefore have the law of conservation of angular momentum.

The term ignorable coordinate means just what is say- that for corresponding p_α simply as a constant appearing in the Hamiltonian function. Because of conservation law for angular momentum, we were able to deal with an effectively one-dimensional problem involving only the radial coordinate r. The generalized momentum $p_\theta = J$ was simply a constant appearing in the equation of motion or the energy conservation equation.

Let us reexamine this problem from the Hamiltonian of view. Since the Hamiltonian is independent of θ, θ is ignorable. Thus we may regard 9.76 as the Hamiltonian for a system of one degree of freedom, described by the coordinate r, in which a constant p_θ appears. It is identical with the Hamiltonian for a particle moving in one dimension under a conservative force with potential energy function.

$$U(r) = \frac{p_\theta^2}{2mr^2} + V(r). \tag{9.83}$$

This is the effective potential energy function.

Hamiltonian equations for r and p_r are

$$\dot{r} = \frac{\partial H}{\partial p_r} = \frac{p_r}{m}, -\dot{p}_r = \frac{\partial H}{\partial r} = \frac{dU}{dr}. \tag{9.84}$$

To solve the central force problem, we solve this one-dimensional problem. Our solution gives us complete information about the radial motion-it gives $\dot{r}$ as a function of r as a function of t, by integrating.

Any required information about the angular part of the motion can then be found from the remaining pair of Hamiltonian's equations, one of which is the angular momentum conservation equation $\dot{p}_\theta = 0$, while the other gives $\dot{\theta}$ in terms of p_θ, $\dot{\theta} = \frac{p_\theta}{mr^2}$.

9.11   Liouville's theorem

Liouville's theorem really belongs to statistical mechanics, but it is interesting to consider it here because it is a very direct consequences of Hamilton's equations.

The instantaneous position and velocity of every particle in our system is specified by the $2n$ variables $(q_1, q_2, \cdots, q_n, p_1, p_2, \cdots, p_n)$. It is convenient to think of these coordinates in a $2n$ dimensional space, called the phase space of the system. Symbolically we may write

$$\mathbf{r} = (q_1, q_2, \cdots, q_n, p_1, p_2, \cdots, p_n) \qquad (9.85)$$

As time progresses, the changing state of the system can be described by a curve $\mathbf{r}(t)$ in the phase space. Hamilton's equations prescribe the rates of change $(\dot{q}_1, \cdots, \dot{q}_n, \dot{p}_1, \cdots, \dot{p}_n)$. These may be regarded as the components of a $2n$ dimensional velocity vector,

$$\mathbf{v} = (\dot{q}_1, \cdots, \dot{q}_n, \dot{p}_1, \cdots, \dot{p}_n). \qquad (9.86)$$

Now suppose that we have a large number of copies of our system, starting out with slightly different initial values of the coordinates and momenta. For example, we may repeat many times an experiment on the same system, but with small random variations in the initial conditions. Each copy of the system is represented by a point in the phase space, moving according to Hamilton's equation. We thus have a swarm of points, occupying some volume in phase space, rather like the particles in a fluid.

Liouville's theorem concerns how this swarm moves. What is says is a very simple, but very remarkable result, namely that the representative points in phase space move as though they formed an incompressible fluid. The $2n$ dimensional volume occupied by the swarm does not change with time, though its shape may change in very complicated ways.

To prove this, we need to apply the generalization of the divergence operation. In three dimensions, it is shown that the fluid velocity in an incompressible fluid satisfies the condition $\nabla \cdot \mathbf{v}$. It is easy to see that the argument generalizes immediately to any number of dimensions. The condition that the phase-space volume does not change in the flow described by the velocity field $\mathbf{v}$ is simply that the $2n$-dimensional divergence of $\mathbf{v}$ is zero,

$$\nabla \cdot \mathbf{v} = \frac{\partial \dot{q}_1}{\partial q_1} + \cdots + \frac{\partial \dot{q}_n}{\partial q_n} + \frac{\partial \dot{p}_1}{\partial p_1} + \cdots + \frac{\partial \dot{p}_n}{\partial p_n}. \tag{9.87}$$

But by 9.64 and 9.65, the first and $(n+1)$ terms are

$$\frac{\partial \dot{q}_1}{\partial q_1} + \frac{\partial \dot{p}_1}{\partial p_1} = \frac{\partial}{\partial q_1}\left(\frac{\partial H}{\partial p_1}\right) + \frac{\partial}{\partial p_1}\left(-\frac{\partial H}{\partial q_1}\right). \tag{9.88}$$

which is indeed zero. Similarly all the other terms in 9.87 cancel in pairs. Thus the theorem is proved.

In general except for special cases, the motion in phase space is complicated. The phase-space volume containing the swarm of representative points maintains its volume, but becomes extremely distorted, rather like a drop of immiscible coloured liquid in a glass of water which is then stirred.

The points cannot go anywhere in phase space, because of the energy conservation equation. They must remain on the same constant-energy surface, and of course there might be other conservation laws that also restrict the accessible region of phase space. But one might expect that in time the phase-space volume would become thinly distributed throughout almost all the accessible parts of phase space. Such behavior is called "ergodic", and is commonly assumed in statistical mechanics. Averaged properties of the system over a long time can then be estimated by averaging over the accessible phase space.

Remarkably, however many quite complicated systems do not behave in this way, but show surprising almost periodic behavior. The study of which systems do an do not behave ergodically, and particularly the transition between one type of behavior and another as the parameters of the system are varied, is now one of the most active fields of mathematical physics. It has revealed an astonishing richness of possibilities.

Chapter 10

Special relativity

10.1 Inertial frame

The idea of an inertial frame of reference or just "inertial frame" is one that already plays an important role in non-relativistic mechanics. It is an attempt to formalize the notion that motion is relative in an operational way. To do this, we must carefully describe what velocity means by describing precisely what we need to measure it.

On the surface, the speed of light does not seem to be a complicated concept. You measure it in the obvious way with clocks and meter sticks, by dividing the distance traveled by the time taken. But first, you have to synchronize your clocks. This is where the idea of an inertial frame comes in. An inertial frame is a real or imaginary collection of clocks that are fixed with respect to one another and synchronized, for example by requiring that some signal that originates midway between each pair of clocks arrives at the two clocks at the same time. In addition, an inertial frame must not be accelerating, which is easy to check because you can just demand that

Newton's laws hold for small velocities-free particles travel in straight lines, that sort of thing.

So you have two fundamental principles.

A. That the laws of physics are the same in all inertial frames.

B. That one of the laws of physics is that the velocity of light is a constant- with the same value in all inertial frames.

As you will see in more detail, these two principles are amazingly powerful. They will revolutionize our picture of space and time.

Speed of light

The speed of light is exactly 299,792,458 m/s. Because the speed of light is built into structure of space and time, it makes sense to use it to define our unit of length in terms of our unit of time. This is what is done in SI units. It is no longer necessary to keep a standard 1 meter bar in a vault someplace. The second is now define in terms of a particular oscillation of an atom in an atomic clock. The meter is then defined as the distance that light travels in 1/299792458 th of a second. We should say that when we talk about the speed of light, it always mean the speed of light in vacuume-that is in empty space. Things get more complicated in material like glass because the interactions of the light with the material can slow the light down.

Now 299792458 m/s is fast. It is a heck of a lot faster than we can actually move ourselves. But it is certainly not infinitely fast. With modern electronics, we can measure very short times, so it is not impossible to see the effect of the finite speed of light even over fairly short distances. The point we are trying to make here is that while motion at close to the speed of light is far beyond our everyday experiences, it is not science fiction. In fact we routinely measure the speed of light, and routinely see things moving at speeds very close to the speed of light.

But the surprising thing about light in a vacuum is that the speed of light that we measure doesn't depend on the velocity of the object that produced the light, and it doesn't depend on the velocity of the measuring apparatus. If, for example, I am running towards a light-bulb at speed v carrying a light-speed mater, a device to measure the speed of light, all of you sitting at rest see the light from the bulb approaching me at a speed v+c. But when I do the measurement, I get the same value for the speed of light that I would get if I were standing still. In fact, I get the same value that you would get measuring the same light beam in about the same place at about the same time, but standing still. The same thing happens if I am running away from the light source. If I am moving towards the light beam, I should register a larger speed on my light-speed meter. That is what common sense would say. However that is not the way the world works. The way the world works is that the speed of light in vacuum is constant. It is not that something goes wrong with my light-speed meter. This bizarre fact is built into the way the world works.

The full power of this remarkable fact, the constancy of the speed of light, is unleashed when we combine it with another, much more reasonable fact about the way the world works-the principle of relativity. The principle of relativity says simply that all uniform motion is relative. There is no absolute sense in which I can say I am moving. There is no preferred notion of standing still. In a moment, we will formalize this idea with the notion of an inertial frame of reference. Note that we can tell if our motion is not uniform. Accerelation is always accompanied by forces that we can feel in our bones. But uniform motion is not detectable, so long as everything else we nee is moving along with us. This is something that we feel in our bones for the slow motion that we are used to. We all know this very well from travel in vehicles, cars, trains, planes. We are going to assume that it remains true at ralativistic speeds.

10.2   Time dilation

Let's start with one of the strangest and most trivial of the consequences of relativity-time dilation. The phenomena of time dilation can be stated precisely as follows. Observation done on a single clock moving with speed v with respect to a number of clocks fixed in an inertial frame show the ticking of the moving clock slowed down by a factor $\sqrt{1 - \frac{v^2}{c^2}}$. The standard way of deriving this result is to consider an idealized clock made out of two parallel mirrors and a pulse of light bouncing back and forth between them:

If the distance between the mirrors is L, the time for each tick of the clock, defined as the time for the pulse to get from one mirror to the other, is L/c. Now suppose that the two mirrors of this light clock are mounted on parallel tracks a distance L apart and the two mirrors moved down the tracks with velocity v.

Obviously, from the point of view of the many clocks in the inertial frame, the light pulse has to go farther when the single light clock is moving. Thus if light always travel at the same velocity, the ticks of the light clock take longer when it is moving. Call the factor by which the ticks are longer γ. Then we can compute γ as follows. Each vertical transit of the light from one mirror to the other in the moving frame takes time $\gamma L/c$. And because light travels at the same speed, c, the length of the path from one mirror to the next is therefore $\gamma L/c$. The light pulse moves vertically a distance L and horizontally a distance $v\hat{\gamma}L/c$. Now we look at the geometry of the motion. Then Pythagoras tells us that

$$(\gamma L)^2 = (\gamma Lv/c)^2 + L^2. \tag{10.1}$$

which implies

$$\gamma = \frac{1}{\sqrt{1 - \frac{v^2}{c^2}}} \tag{10.2}$$

So time is no longer sacred. And this can't just be a special property of light clocks, because if we used some other kind of clock to measure the time, and got a different result, then we would be able to distinguish between the moving frame and the frame in which the light clock is fixed. But this violates the principle that all frames are equivalent. Every kind of clock must tick out seconds at the same rate in all inertial frames.

Incidently this factor γ is going to reappear all the time, so it pays to actually either memorize it, or to be able to reproduce the light-clock argument in real time so you can get it whenever you need it.

It is quite easy with modern electronics and atomic clocks to see relativistic effects like time dilation. In fact, both special and general relativistic effects are very important in one very practical application-the Global Positioning System which is based on a system of atomic clocks aboard satellites. The relativistic corrections are small, because the satellites are travelling at only about 4000m/s, but enormous accuracy is required to make GPS work and the relativistic effects must be properly included.

Even more dramatic examples of time dilation occur all the time with elementary particles. That seems like a lot to ask of tiny particles that are supposed to be elementary and have no internal structure. But the fact is that quantum mechanics provides us with internal clocks for many elementary particles because they are unstable, and when they are sitting still and evolving in time, they have a constant probability per unit time of decaying into other lighter particles. We can actually see these internal clocks ticking by watching the particles decay. The observed lifetime of unstable particles is a tangible measure of how fast these internal clocks are ticking. We see this all the time in particle experiments. But we are relying on another fact-all particles of a particular kind are exactly the same. We never actually measure the decay rate of the same decaying particle in two different frames. But we can quite easily measure the lifetime of a particle at rest, and then measure the lifetime of the same type of particle in a moving frame. We find the ratio of the lifetimes is γ. Since all particles of a particular type are identical, this is just as good.

Twin paradox

Now you may very well be thinking that once you define what you are thinking about carefully, with inertial frame, that there isn't anything particularly strange about motion at relativistic speeds, but that we have just confused the issue with a bizarre definition of measurement. Even our experiment on decays of elementary particles might be just a matter of a bad definition of what we mean by the ticking of their inertial clocks. Perhaps that there is some other way of constructing our light-speed meters so that the speed of light is not constant and the bizarre features of relativity go away. Think again. Perhaps the simplest way of making clear that something totally bizarre is going on is to discuss the twin paradox. This is a classic thought experiment in which one twin takes a trip on a rocket moving at relativistic speeds, while the other twin remains at home. When the tarveling twin returns, because his clocks have been ticking slowly, he is younger than his twin. His biological clock is no different from any other clock. Relativity has slowed down the aging process. If this is not strange, I don't know what is.

Again, this experiment has not been done conclusively with people. Astronauts in MIR or the space shuttle do age less rapidly than the rest of us, but the difference is sufficiently small at the speeds of mere orbital motion that the don't see a huge difference in biological clock. The difference can be measured atomic clocks. And a number of very accurate experiments have been done showing exactly this effect with the internal clocks of unstable elementary particles.

The twin paradox is so peculiar that I want to work out an example of how it looks to the two twins who are aging differently. To do this, it is useful to first understand the relativistic Doppler effect.

Doppler effect

While the speed of a light beam does not change when we go from one inertial frame to another, its frequency does change. This is not surprising, since the same thing happens for sound or any other wave. It is called the Doppler effect and shows up, for example, in the change in sound of a train whistle when the train goes by. First, let me remind you how the Doppler effect goes for sound, or any other wave when we are moving at nonrelativistic speeds. Suppose that a train is moving towards me at speed v and its whistle emits sound waves which, when the train is at rest, have frequency $/nu_0$, and wavelength λ_0. The speed of sound V_s, is the product of the frequency and the wavelength:

$$V_s(m/s) = \lambda_0(m/cycle)\dot{\nu}_0(cycles/s).$$ (10.3)

The inverse of the frequency is the period of the sound wave, which is the time between successive crests of the wave. So now let us look at two successive crests of the wave as the train moves towards us.

Because the train is moving forward as it emits the wave, the crests are closer together than they would be if the train were standing still. The distance between crests for the train at rest is just the wavelength. The distance between crests for the moving train is

$$\lambda_v = \frac{V_s - v}{\nu_0}$$ (10.4)

Thus the wavelength of the sound as recorded at the sound meter is reduced by the nonrelativistic Doppler factor.

$$\frac{V_s - v}{V_s}.$$ (10.5)

Because 10.9 must be satisfied, the frequency is increased by the inverse of ??, and the train whistle has a higher pitch when it is moving towards us.

If the train is moving away, the argument is exactly the same, we just replace $v \rightarrow -v.$

Now suppose we do a similar thing, but replace the train with a rocket moving at relativistic speed, and replace sound with light. The speed of light is the product of the frequency and the wavelength:

$$c(m/s) = \lambda_0(m/cycle)\dot\nu_0(cycles/s) \tag{10.6}$$

The inverse of the frequency is the period of the light wave, which is the time between successive crests of the wave. In the diagram, almost everything is the same, except that because of time dilation, the time between the emission of successive crests of the wave is longer than $\frac{1}{\nu_0}$ by the ubiquitous factor of γ, because the moving clock ticks more slowly.

The distance between crests for the rocket at rest is just the wavelength, $\lambda_0 = \frac{c}{\nu_0}$.

Thus the wavelength of the light as recorded at the light meter is reduced by the relativistic Doppler factor

$$\frac{\gamma(c-v)}{c} = \frac{1}{\sqrt{1 - \frac{v^2}{c^2}}}(1 - \frac{v}{c}) = \sqrt{\frac{1 - v/c}{1 + v/c}}. \tag{10.7}$$

and the frequency is increased by $\sqrt{\frac{1+v/c}{1-v/c}}$, and the light has higher frequency when the rocket is moving towards us. This is called blue-shift because raising frequency in the optical spectrum is a shift towards the blue.

Again, if the rocket is moving away, the argument is exactly the same- we just have to replace $v \to -v$ everywhere. This is called red-shift because lowering frequency in the optical spectrum is a shift towards the red.

There is one very important distinction to note about the relativistic Doppler effect versus the nonrelativistic version. In the relativistic version, it doesn't matter whether the rocket is approaching the observer at speed v or the observer is approaching the rocket at speed v. It can't, because of the principle of relativity. This is not true for the nonrelativistic Doppler effect because the air in which sound moves difines a special frame.

10.3 Lorentz contraction

Lorentz contraction is the statement that moving observers see stationary objects contracted(in the direction of motion) by a factor $\frac{1}{\gamma} = \sqrt{1 - \frac{v^2}{c^2}}$. We have already seen that a single clock fixed in a frame moving with speed v appears to tick slowly in our frame by a factor of $\frac{1}{\gamma}$. Suppose this clock moves past a measuring stick fixed in our frame with length L. The time this takes in our frame is $\frac{L}{v}$. Thus the time it takes according to the fixed clock in the moving frame is $\frac{L}{\gamma v}$. But the moving observer sees the measuring stick moving with speed v. Thus he concludes that since it took a time $\frac{L}{\gamma v}$ for the stick to get past him, the stick has length $\frac{L}{\gamma}$. We will come back later to a more detailed discussion of exactly what this means, but it has to work this way.

Relativity of sumulatneity

The classic example of relativity of simultaneity is the case of two clocks at opposite ends of a train car as the train moves through a station at speed v. If these clocks are synchronized according to observers on the train car, then light signals going forward and backward from the exact center of the train car will reach the two clocks when they read the same time. In fact, this would be a good way to synchronize the clocks in the first place. But when this process is viewed by observers in the station, the backward-going light reaches the rear clock before the forward-going light reaches the front clock. Thus to observers in the station, when the light reaches the rear of the train, the clock in the front of the train must be reading an earlier time, because the light has not gotten there. We summarize this peculiar situation by saying that it is earlier in the front of the train. The word front indicates that we are refering to a frame in which the train is moving and fixes the direction. It is important in dealing with relativity to be alert for hidden clues like this. More precisely, what this means is that clocks synchronized in the train frame have different readings at the same time in the station frame, with the clocks in front having the earlier reading.

Moving light clock

Let me illustrate some of these issues by talking about a particularly useful sort of spacetime event-the tick of a clock. If I have a clock sitting at the origin of the reference frame, ticking every T seconds, the ticks of the clock define a series of spacetime events. In the frame of reference, all of these events have space coordinates 0, because they are all at the origin, and their time coordinates are just 0, T, 2T, etc. If I focus on the x and t coordinates for simplicity, the series of ticks would look like

$$(t, x) = (0,0), (T,0), (2T,0) \cdots \tag{10.8}$$

In general, if we hadn't put the first tick at the origin, the x and t coordinates of the later ticks would all just have the initial x and t positions added on to them, so the sequence would look like

$$(t, x) = (t_0, x_0), (t_0 + T, x_0), (t_0 + 2T, x_0), \cdots \qquad (10.9)$$

We know that there is nothing very interesting about the x_0 and the t_0, which can be changed by moving to a different coordinates system by translations in space and time.

But we also know from our discussion of time dilation what these same events look like in a reference frame in which the clock is moving. Suppose that the clock is moving with velocity v in the +x direction. Because of time dilation, the ticks are spread out by a factor of γ, so that the time coordinates are $0, \gamma T, 2\gamma T$, etc. Let us also assume that the x coordinate is 0 for the first tick (at t=0). Then because we know that velocity of the clock, we can easily work out the other x coordinates

$$(t, x) = (0, 0), (\gamma T, v\gamma T), (2\gamma T, 2v\gamma T), \cdots \qquad (10.10)$$

Again, if we hadn't put the first tick at the origin in space and time, the x and t coordinates of the later ticks would all just have the initial x and t positions added on to them, so the sequence would look like

$$(t, x) = (t_0', x_0'), (t_0' + \gamma T, x_0' + v\gamma T), (t_0' + 2\gamma T, x_0' + 2v\gamma T), \cdots \qquad (10.11)$$

While you are used to thinking of the difference between 10.8 and 10.9 as a change of coordinate system, you are probably not used to thinking about the difference between 10.8 and 10.11 as a change of coordinate system. But what you see here is that these spacetime coordinates really can describe the same set of events, the ticks of a clock, in two different reference frames. Going from one reference frame to another, then is like going from one coordinate system to another-it is just that the time coordinate as well as the space coordinates are involved. What this means is that we have gone from a 3 dimensional space to a 4 dimensional spacetime.

The invariant interval

We can extend the analogy between going from one coordinate system to another and going from one reference frame to another in the following way. When you make a chance of coordinate system in space, the distance between points doesn't change. You can easily compute the square of the distance between two points labeled by vectors $\vec{r}_1$ and $\vec{r}_2$ in terms of their coordinates,

$$l^2 \equiv (\mathbf{r}_1 - \mathbf{r}_2) \cdot (\mathbf{r}_1 - \mathbf{r}_2) = \Sigma_{j=1}^{3}(\mathbf{r}_1^j - \mathbf{r}_2^j)^2. \tag{10.12}$$

Two things have happened here. First we have subtracted the coordinates of the two points. This takes care of possible translations, which just cancel when we subtract. Another way of putting this is that we are not interested in the length of the vectors themselves, because this depends on the arbitary position of the origin. But if we make the vector from $\mathbf{r}_2$ to $\mathbf{r}_1$ by forming the combination $\mathbf{r}_1 - \mathbf{r}_2$, it doesn't depend on the origin. Then we sum the squares of the coordinates to get something that doesn't change when we make a rotation.

When you make a chance of coordinate system in spacetime, by going to a new reference frame, there is a similar thing that doesn't change. If I have two events, event 1 at time t_1 and position $\mathbf{r}_1$ and event 2 at time t_2 and position $\mathbf{r}_2$, the following combination doesn't change when we go from one reference frame to another.

$$s^2 \equiv c^2(t_1 - t_2)^2 - (\mathbf{r}_1 - \mathbf{r}_2) \cdot (\mathbf{r}_1 - \mathbf{r}_2) \tag{10.13}$$

doesn't change when we go from one reference frame to another. Again we have done two things. The obvious one is to subtract the space and time coordinates of the two events, so that s^2 doesn't depend on the origin of coordinates. The other, much less obvious one, is to combine the difference in time and the difference in space in a very particular way. You can see that this is the right thing to do by looking at two successive clock ticks. We have

$$s^2 = c^2 T^2 - 0^2 = c^2 T^2 \tag{10.14}$$

and

$$s^2 = c^2(\gamma T)^2 - (v\gamma T)^2 = (c^2 - v^2)T^2/(1 - v^2/c^2) = c^2 T^2 \tag{10.15}$$

This is independent of v, so it looks like the same in any frame of reference.

The quantity s^2 looks like the quantity l^2. There are three differences. One is that time is involved. The second is the minus sign in front of the second term. The third is the factor of c^2 in the first term.

Let's deal with the easy one first. The factor c^2 is nature's way of telling us that we are using a really stupid system of units. Because time and space get mixed up in relativity, and because the ratio v/c appears in many of the relations of relativity, it makes sense to use units in which $c = 1$. One way of doing this is to use seconds as your unit of time and light-seconds, that is the distance light travels in a second, 299,792,458 meters. Of course, because we are so slow, this is an inconveniently long distance but you will get used to it. Anyway, 1 is easier to remember than 2999,792,458. So from now on, we will follow nature's advice and set c=1, which we will call relativistic units. This will make our formulas look simpler. We also something by doing this, but it is not very important.

The other two differences between s^2 and l^2 express the crucial strangeness of relativity. Time is not quite the same as space-that is obvious from the minus sign. But it is not completely different either, because it must be included to get something that looks the same in different reference frames.

Notice that because of the minus sign, there is no reason why s^2 has to be positive. It is positive in our example of the interval between two ticks of a clock, but for other kinds of intervals, such as the interval between two events that are in different places but at the same time in some frame, s^2 can be negative as well. The quantity s^2 is called invariant interval.

10.4 Lorentz transformations

You have read about a nice derivation of Lerentz transformation. They look much simpler with $c = 1$. So suppose that we have two events

event 1 at $t_1, \mathbf{r}_1$ and

event 2 at $t_2, \mathbf{r}_2$.

The interesting thing is the interval in time and space between these two events:

$$\Delta t = t_2 - t_1, \Delta \mathbf{r} = \mathbf{r}_2 - \mathbf{r}_1. \tag{10.16}$$

Now if we look at these same two events from a reference frame moving in the $+x$ direction with speed v, the coordinates of the events will change. We will get a new description of the events in terms of new coordinates,

event 1 at $t_1', \mathbf{r}_1'$ and

event 2 at $t_2', \mathbf{r}_2'$.

Again, the interesting thing is the interval in time and space between these two events:

$$\Delta t' = t_2' - t_1', \Delta \mathbf{r}' = \mathbf{r}_2' - \mathbf{r}_1'. \tag{10.17}$$

Now the statement of the Lorentz transformation is that we find that we can choose a coordinate system in which

$$\Delta x' = \gamma(\Delta x - v\Delta t), \tag{10.18}$$

$$\Delta t' = \gamma(\Delta x - v\Delta x), \tag{10.19}$$

$$\Delta y' = \Delta y, \Delta z' = \Delta z, \tag{10.20}$$

If we set $\Delta \mathbf{r} = 0$ and look at $\Delta x'$ and $\Delta t'$ as a function of Δt, we can recognize the moving clock. The condition $\Delta \mathbf{r} = 0$ means that we are looking at a series of events that are fixed in the original frame. A pair of events with Δt could be two ticks of a fixed clock a time Δt apart.

$$\Delta \vec{r} = 0 \tag{10.21}$$

$$\Delta x' = \gamma(-v\Delta t), \tag{10.22}$$

$$\Delta t' = \gamma(\Delta t), \tag{10.23}$$

$$\Delta y' = 0, \Delta z' = 0 \tag{10.24}$$

The γ in

$$\Delta t' = \gamma(\Delta t) \tag{10.25}$$

expresses time dilation.

The $-v$ in

$$\Delta x' = \gamma(-v\Delta t). \tag{10.26}$$

describes the clock moving with x velocity $-v$ in the new frame.

The other term in the Lorentz transformation can be understood in a similar way.

It is important to understand why Lorentz transformation exist. They are a kind of space-time analog of rotations, but they look different because of the minus sign.

Addition of velocities

There is one case in which Lorentz transformations are probably actually the simplest way to proceed. That is the problem of "addition of velocities", which is the question of how motion in one inertial frame looks in another inertial frame. Let us first consider the relatively simple case in which the motion in question and the motion of the inertial frame are in the same direction. In particular consider two events on the path of particle moving with velocity u in the $+x$ direction in our inertial frame. Suppose that the time difference between the events in our frame is $\Delta t = T$. Then the x interval between them is $\Delta x = uT$ so that the x component of velocity is $u_x = \Delta x / \Delta t = u$. The y and z interval are zero. Now what happens when we look at these same two events in an inertial frame moving with velocity v in the x direction. If this were non-relativistic physics, we would expect that the new intervals in the moving frame would describe motion with velocity $u - v$. But we know that something weird is going to happen in relativistic physics. We already know from the fundamental postulate about the speed of light that if $u = c$, that is the original motion is at the speed of light, the motion in the new inertial frame must also be at the speed of light. Let's see how this works. We can read off the intervals in the new frame using the Lorentz transformation:

$$\Delta x' = \gamma(\Delta x - v\Delta t) = \gamma(uT - vT), \tag{10.27}$$

$$\Delta t' = \gamma(\Delta t - v\Delta x) = \gamma(T - vuT), \tag{10.28}$$

$$\Delta y' = \Delta y = 0, \Delta z' = \Delta z = 0. \tag{10.29}$$

Using the above, we can read off the velocity in the new frame just by dividing:

$$\frac{\Delta x'}{\Delta t'} = \frac{\gamma(uT - vT)}{\gamma(T - vuT)} = \frac{u - v}{1 - uv}. \tag{10.30}$$

So you see that if u and v are much smaller than c , the second term in the denominator of (21) is negligible and we recover our non-relativistic expectation, $u - v$. On the other hand , if $u = c$, we get

$$\frac{\Delta x'}{\Delta t'} = \frac{c - v}{1 - cv/c^2} = \frac{c - v}{1 - v/c} = \frac{c - v}{(c - v)/c} = c, \tag{10.31}$$

as we must. So while (20) doesn't make any sense to us, it is perfectly consistent, and in fact it is right.

Note that if u and v have opposite signs, so that observers in the moving frame are approaching the moving particle, that the denomenator in (20) becomes greater than one. This prevents the motion in the moving frame from ever appearing to be faster than light.

Relativistic units

So how did we know how to put the factors of c in to 10.125? This is a crucial step because it allows you to do calculations with c=1, to make things simple, but still get the full answer at the end of the day, so you can talk to people who insist on using dumb units. If you have done a calculation in relativistic units and you want to put the factors of c back into translate the result to conventional units, you have to know the conventional units of the objects you have calculated. Then you put in factors of c to get the units right.

In the case of 10.125, we know that we are calculating a velocity, and the numerator is already a velocity. So we can leave that alone and make the denomenator dimensionless. The 1 term is OK, and we can make the uv term dimensionless by dividing by two factors of c.

Suppose that you are calculating an energy in terms of a mass m and a velocity v and you get the result $6\pi mv^4$. The 6π is a dimensionless number- we can forget about that. But the rest would not have units of energy in conventional units. We have two too many factors of v, so we must divide by c^2. Thus the result in normal units is

$$6\pi mv^4/c^2 \tag{10.32}$$

Here's another one. For a system of two particles, with masss m_1 and m_2 and speeds v_1 and v_2, suppose you calculate an object that is a mass, and the result is

$$\frac{m_1/v_1 + m_2v_2}{m_1/m_2 + v_1v_2} \tag{10.33}$$

Let us start with the denominator. We don't know what the units should be of the denominator alone, but it is clear that the two terms in the denomenator must have the same dimension. So for example, we can divide the second term by c^2. That makes the whole denominator dimensionless, so the numerator must then have the dimensions of mass. We can arrange this by multiplying the first term by c and dividing the second by c-thus the result in normal units is

$$\frac{m_1c/v_1 + m_2v_2/c}{m_1/m_2 + v_1v_2/c^2} \tag{10.34}$$

$\hbar = 1$ and the Plank scale

As we learn more about relativity we will come more and more to appreciate the speed of light c as a kind of cosmic speed limit, built into the fabric of space-time. There is at least one and perhaps two other such constants. The obvious one is the fundamental constant of action in quantum mechanics, $\hbar$, or Plank's constant over 2π. Particle physicists usually not only set c=1, but also $\hbar = 1$ as well. $\hbar$ has the units of angular momentum, mass$\times$ distance $\times$ velocity, or momentum$\times$ distance. We can call a set of units in which $c = \hbar = 1$ particle physics units. In particle physics units, everything can be measured in powers of energy or length or momentum of time, whatever you like, and you can go back and forth. As with relativistic units, if you remember what it is you are calculating, you can do your calculations in particle physics units so you don't have to carry around these constants, and always go back and put the c and $\hbar$ in at the end of the calculation. Here are some examples.

1. The fine structure constant : In electrostatic units, Coulombs's law for the magnitude of the force between two charges q_1 and q_2 is

$$\frac{q_1 q_2}{r^2} \tag{10.35}$$

The charges are measured in electrostatic units(esu) defined as the charge that produces a force of 1 dyne at a distance of 1 centimeter. In esu, the charge of the electron is

$$e \approx 4.80 \times 10^{-10} esu \tag{10.36}$$

since e^2/r^2 is a force, which is a mass times an acceleration, e^2 has units of mass time velocity squared distance, which is angular momentum times velocity. Thus the combination

$$\alpha \equiv \frac{e^2}{\hbar c} \tag{10.37}$$

is dimensionless. In cgs units the constants are approximately

$$c \sim 2.998 \times 10^{10} cm/s, \hbar \approx 1.054 \times 10^{-27} dynes \tag{10.38}$$

Thus

$$\alpha \approx \frac{e^2}{\hbar c} \approx \frac{1}{137.15} \tag{10.39}$$

For a while, it was believed that α was exactly $1/137$. This led to a lot of silly numerology, some even by otherwise respectable people. But what is really important about is that α is small. This means that the electromagnetic force is rather feeble in an absolute sense. It would be impossible to make a statement like this if the value of the charge of the electron depended on the units we used to measure it. But once we recognize that c and $\hbar$ are built into the design of the universe and set them equal to one, there is no further such dependence. While the value of e depends on units, the fine structure constant α doesn't. The smallness of α was very important historically, because it allowed physicists to do accurate calculations of the properties of this force in a power series expansion in powers of α. This was worked out by Richard Feynman and Julian Schwinger.

2. The electron Compton wavelength and the Bohr radius: The mass of the electron is

$$m_e \approx 9.11 \times 10^{-28} g \tag{10.40}$$

We can turn $1/m_e$ into a distance as follows:

$$\lambda_e = \frac{\hbar}{m_e c} \approx 3.86 \times 10^{-11} cm \qquad (10.41)$$

The size of atoms depends both on the mass of the electron, and on the strength of the electomagnetic force. The quantum mechanical analysis of the hydrogen atom shows that its size of order

$$R_{Boh} = \frac{\lambda_e}{\alpha} \approx 0.529 \times 10^{-28} cm \qquad (10.42)$$

about half an Angstrom. This is called the Bohr radius.

3. The Plank mass: The other constant besides c and $\hbar$ that may be built into the structure of the universe is Newton's constant, G_N that describes the strength of the gravitational force between masses m_1 and m_2 at a distance r

$$\frac{G_N m_1 m_2}{r^2} \qquad (10.43)$$

where

$$G_N \approx 6.67 \times 10^{-8} cm^3 g^{-1} s^{-2} \qquad (10.44)$$

$G_N m^2$ has the same units as e^2, so we can compare the strength of the gravitational force between electrons to the electric force by looking at the ratio

$$\frac{G_N m_e^2}{e^2} \approx 2.4 \times 10^{-43} \qquad (10.45)$$

Gravity is an absurdly, ridiculously weak force. We see it only because you cannot neutralize it. To make a mass out of G_N we can form the combination

$$M_{Plank} = \sqrt{\hbar c / G_N} \approx 2.18 \times 10^{-5} g \qquad (10.46)$$

This is not a big mass, but it is an absurdly huge mass for an elementary particle:

$$\frac{M_{Plank}}{m_e} \approx 2.4 \times 10^{22} \qquad (10.47)$$

The fact that gravity is so weak, or alternatively that the Plank scale is so huge, a remakable mystery about the world.

Varieties of space-time intervals

We spent a lot of time talking about the space-time interval $(\Delta t, \Delta \mathbf{r})$ between two ticks of a moving clock. This is called a time-like interval, because there is a frame in which it has a time component but no space component- the frame in which the clock is at rest. In fact, any space and time interval with components

$$(\Delta t, \Delta \mathbf{r}) \qquad (10.48)$$

that satisfies

$$\Delta t > |\Delta \mathbf{r}| \qquad (10.49)$$

has the property that there is a frame in which it looks like $(\Delta \tau, 0)$ with $\Delta \tau > 0$. Why? One way to see this is to note that any interval that satisfies 10.48 could be the interval associated with a clock moving with velocity

$$\mathbf{v} = \frac{\Delta \mathbf{r}}{\Delta t} \qquad (10.50)$$

simply because if the clock moves $\mathbf{v}$ for a time Δt, it ends up translated by the vector $\Delta \mathbf{r}$ reproducing 10.48. Thus if we go to a frame moving with velocity $\mathbf{v}$, we will be moving along with the clock and in this new frame the intervals will have the form

$$(\Delta \tau, 0) \tag{10.51}$$

,where $\Delta \tau = \sqrt{(\Delta t)^2 - |\Delta \mathbf{r}|^2}$.

This relation incorporates the statement of time dilation. $\Delta \tau$ is always less than or equal to Δt.

Two events that are separated by a time-like interval are closest together in time in the frame in which they occur at the same point in space.

The frame in which the two events occur at the same point in space is the frame in which the two events could be two ticks of the same clock. Thus the above statement is equivalent to the statements referred to as time-like separated.

Space-like intervals and Lorentz contraction

On the other hand, if

$$\Delta t < |\Delta \mathbf{r}| \tag{10.52}$$

then it cannot represent the interval between two points on the path of a clock. As we will see in more detail later, we cannot go to a frame moving with velocity $\mathbf{v} = \frac{\Delta \mathbf{r}}{\Delta t}$, because $|\mathbf{v}| > 1$, so the frame would have to moving at a speed greater than the speed of light, which is impossible for clocks and meter sticks and all the other things we need to have a frame of reference. Then what is this interval? Suppose that we look in a frame of reference moving with velocity

$$\mathbf{u} = \Delta\mathbf{r}\frac{\Delta t}{|\Delta\mathbf{r}|^2} \tag{10.53}$$

To see what happens, let us rotate our coordinate system until $\Delta\mathbf{r}$ is in the x direction

$$\Delta\mathbf{r} = |\Delta\mathbf{r}|\hat{r} \tag{10.54}$$

so that

$$\mathbf{u} = u\hat{x} \tag{10.55}$$

,where $u = \frac{\Delta t}{|\Delta\mathbf{r}|}$.

Now in a frame moving with velocity $\mathbf{u}$, the y and z components of the interval remain zero, and we can compute what the time and x components become using the Lorentz transformation,

$$\Delta t' = \frac{1}{\sqrt{1 - u^2}}(\Delta t - u|\Delta\mathbf{r}|) = 0. \tag{10.56}$$

$$\Delta\mathbf{r}' = \frac{1}{\sqrt{1 - u^2}}(|\Delta\mathbf{r}| - u\Delta t) = \frac{1}{\sqrt{1 - u^2}}|\Delta\mathbf{r}|(1 - u\frac{\Delta t}{|\Delta\mathbf{r}|}) \tag{10.57}$$

$$\frac{1}{\sqrt{1 - u^2}}|\Delta\mathbf{r}|(1 - u^2) = \frac{1}{\sqrt{1 - u^2}}|\Delta\mathbf{r}| = \sqrt{|\Delta\mathbf{r}|^2 - (\Delta t)^2}. \tag{10.58}$$

This is called a space-like interval, because there is a frame of reference in which it has a space component but no time component. A very important example of a space-like interval is the interval between two ends of a measuring stick at fixed time. This is what we define to be a measurement of distance. Two objects are a distance one meter apart in a given reference frame at time t if we can line up a meter stick so that the objects are at opposite ends of the stick at time t. This distance is associated with a space-time interval with is the difference between two events which are the space-time coordinates of the two ends of the meter stick at time t. This is obviously a space-like interval, because the time component is zero. This corresponds to the primed frame in 10.58. The properties of space-like intervals are then responsible for the phenomenon of Lorentz contraction. In any other frame, the two events occur at different times. Therefore 10.58 implies that $|\Delta \mathbf{r}| > |\Delta \mathbf{r}'|$. The tow events are closet together in the frame in which they occur at the same time. This is generally true for any space-like interval.

Two events that are separated by a space-like interval are closest together in space in the frame in which they occur at the same time.

Thus for example, if we measure the length of a moving train, the length we measure is the distance between two events describing the positions of the front and back of the train at the same time in our frame. But in the train frame, these two events do not occur at the same time, and thus the distance between them is greater than what we measure. This is Lorentz contraction. Two events that are separated by a space-like interval are sometimes referred to as space-like separated.

World line and proper time

There is a nice relativistic analog of the concept of a trajectory in Newtonian physics- based on the concept of a world line. The idea is that as any massive particle evolves with time, whether it is sitting still or moving, free or accelerating, can always be described by some curve in 4-dimensional space-time that is just the collection of the space-time events $(t, \mathbf{r}(t))$ that describe where the particle is at every time. The reason that it is useful to think about this in this funny 4-dimensional space is that the world line itself has an invariant meaning because it is a collection of space time events that have invariant meaning even though their coordinates will change depending on the inertial frame. The situation is similar to that of a curve in three dimensional space. The curve consists of points that have an invariant meaning, but their coordinates change depending on the coordinate system. Also, just as we can label the points on a curve in three dimensional space by the distance along world line of a particle by measuring the time ticked on the particles internal clock. This is called the proper time, τ. The change in proper time $d\tau$ along any short segment of the world line is just

$$d\tau = \frac{dt}{\gamma} = dt\sqrt{1 - v^2} = \sqrt{dt^2 - d\mathbf{r}^2}, \qquad (10.59)$$

where t and v are measured in whatever coordinate system and inertial frame you like. You can see that this is independent of the inertial frame and the coordinate system by looking at the last term, which is simply the invariant interval for the short line segment. It is crucial here that every short line segment along the world line of a massive particle is time-like. This is true because massive particles can never travel as fast as light.

A good way of describing the world line is to give both t and $\mathbf{r}$ as function of τ:

$$t(\tau)\vec{r}(\tau), \tag{10.60}$$

For example, for a particle at rest

$$t(\tau) = \tau, \mathbf{r}(\tau) = 0. \tag{10.61}$$

and for a particle with velocity $\mathbf{v}$

$$t(\tau) = \gamma\tau, \mathbf{r}(\tau) = \gamma\mathbf{r}\tau. \tag{10.62}$$

10.5   Massive particles

Now let's talk about energy and momentum. We are going to write down the Lagrangian for a free relativistic massive particle. Because this is a free particle, the Euler-Lagrange equation is not all that interesting. We know without thinking about it that it is going to tell us that the particle moves at constant velocity. Nevertheless, finding the right from for the Lagrangian can be instructive. It will allow to ask questions about what would happen if we put in forces, for example. Also, it allows us to identify the quantities that we expect to be conserved because of Noether's theorem and translation invariance in time and space. In fact once we write down these quantities-better known as the energy and the momentum- we will largely forget about the Lagrangian formulation. So don't matter is that you learn to deal with the relativistic energy and momentum that come out of it from Noether's theorem.

What principles should we use to write down the Lagrangian for the free massive particle moving at relativistic speeds? One is that the Action should look the same in all reference frames. This is necessary, because if it were not true, Hamilton's principle would not give us equations of motion that give the same trajectories in all reference frames. In addition, if we call the position of the particle $\mathbf{r}$, we expect that the Lagrangian is independent of $\mathbf{r}$ and of t, and depends only on $\dot{\mathbf{r}}$, because it should be invariant under translations in space and time. In a moment, I will show that the following Lagrangian gives rise to an action that is independent of the reference frame.

$$L(r, \dot{r}) = -m\sqrt{1 - \dot{\mathbf{r}}} \tag{10.63}$$

We are using units with $c = 1$ as usual. Thus looks a bit funny because of the square-root. But as we will see, that is what relativity requires us to write down, so we will just have to live with it. The constant m, which we will see in a moment is the mass of the particle, must be there so that the Lagrangian has units of energy. The same thing happens in the Lagrangian for the free non-relativistic particle, which is just $\frac{1}{2}m\dot{r}$. We can see that m corresponds to the mass of the particle by looking at the Lagrangian for small velocities which we can do by Talyor expansion.

$$-m\sqrt{1 - \dot{\mathbf{r}}} = -m + \frac{1}{2}m\dot{\mathbf{r}} \tag{10.64}$$

The constant -m is irrelavant, so at small velocities, this reduces to the non-relativistic Lagrangian.

Now let us show what happens to the action

$$S = -m \int dt \sqrt{1 - \dot{\mathbf{r}}} \qquad (10.65)$$

under a Lorentz transformation. This is a little complicated because both the integrand and the dt change under a Lorentz transformation. But we can make what is going on more obvious by writing 10.65 as follows

$$S = -m \int dt \sqrt{1 - (d\mathbf{r}/dt)^2} = -m \int dt \sqrt{dt^2 - (d\mathbf{r})^2} \qquad (10.66)$$

The second form is a very funny looking integral,but it makes sense it is equivalent to the previous form. The important point is that the last form is evidently unchanged by Lorentz transformations. The infinitesimal interval

$$(dt, d\mathbf{r}) \qquad (10.67)$$

is just a space-time interval. It transforms under Lorentz transformations just like $(\Delta t, \Delta \mathbf{r})$. And therefore, the combination

$$dt^2 - (d\mathbf{r})^2 \qquad (10.68)$$

is just an infinitesimal version of the invariant interval, and it has the same value in all inertial frames. To get something proportional to dt, we must take the square-root of 10.68. That is why 10.65 looks the way it does.

10.6   Energy and momentum

Now that we have a Lagrangian, we can construct the conserved energy and momentum that we expect for a relativistic particle. The energy is

$$E = \dot{\mathbf{r}} \cdot \frac{\partial L}{\partial \dot{\mathbf{r}}} - L = m\frac{\dot{\mathbf{r}}}{\sqrt{1 - \dot{\mathbf{r}}}} = \frac{m}{\sqrt{1 - \dot{\mathbf{r}}}} = \frac{m}{\sqrt{1 - v^2}} = m\gamma. \qquad (10.69)$$

The momentum is

$$\mathbf{p} = \frac{\partial L}{\partial \dot{\mathbf{r}}} = m\frac{\dot{\mathbf{r}}}{\sqrt{1 - \dot{\mathbf{r}}}} = \frac{m\mathbf{v}}{\sqrt{1 - v^2}} = m\mathbf{v}\gamma. \qquad (10.70)$$

Here is an example. Suppose we have a particle with mass m travelling at $v = \frac{4}{5}$. Then

$$\gamma = \frac{5}{3}. \qquad (10.71)$$

so

$$E = \gamma m = \frac{5}{3}m, p = v\gamma m = \frac{4}{3}m. \qquad (10.72)$$

You will observe that these do not look like the expressions for energy and momentum of a Newtonian particle. Nevertheless, these are the objects that are really conserved. It is instructive to look at them with the factors of c put back in:

$$E = \frac{mc^2}{\sqrt{1 - \frac{v^2}{c^2}}}. \qquad (10.73)$$

$$\vec{p} = \frac{m\mathbf{v}}{\sqrt{1 - \frac{v^2}{c^2}}}. \qquad (10.74)$$

In this form, it is also useful to Taylor expand in powers of $\frac{v}{c}$,

$$E = \frac{mc^2}{\sqrt{1 - \frac{v^2}{c^2}}} = mc^2 + \frac{1}{2}mv^2 + \cdots \tag{10.75}$$

$$\mathbf{p} = \frac{m\mathbf{v}}{\sqrt{1 - \frac{v^2}{c^2}}} = m\mathbf{v} + \cdots . \tag{10.76}$$

As we expected, the Newtonian expressions for the kinetic energy and the momentum appears.

In the expression for energy,(17), there is also a constant term, the famous mc^2. What is actually important about this is not so much the constant mc^2, but the fact that mass play a very different role in relativistic processes than it does at lower speeds. In Newtonian mechanics, mass, momentum and the total energy that are actually conserved in arbitrary collisions of elementary particles at any speed-even in processes in which particles are created or annihilated. Mass is not conserved. The mass of any one kind of particle is always the same, so mass is conserved in collisions that don't change the type of particle involved. For example if you accelerate an electron to speed v and it collides with an electron at rest, some of the time you will get a collision in which you end up with two electrons going off in different directions and no other particles. We might represent this process by the schematic equation

$$e^- + e^- \rightarrow e^- + e^- . \tag{10.77}$$

In this case the sum of the masses of the particles before the collision is the same as the sum of the masses of the particles after the collision, you may produce an extra electron and a positron.

$$e^- + e^- \rightarrow e^- + e^- + e^- + e^+. \tag{10.78}$$

Here the sum of the masses before the collision is $2m_e$, and the sum after the collision is $4m_e$. Mass is not conserved. In fact, there is no reason to compute the sum of the masses at all. It is just not an interesting quantity.

There is a related issue that sometimes causes confusion. Some of you have seen relativity before, and you may have been exposed to the rather idiotic notion of a rest mass that is the actual mass of the particle and a relativistic mass that depends on velocity. This is not useful. If you ask a physicist what the mass of the electron is, the response would be

$$m_e \approx 9.11 \times 10^{-28} g \tag{10.79}$$

or

$$m_e \approx 0.511 MeV. \tag{10.80}$$

Not only do these notions of rest mass and relativistic mass not correspond to the way physicists actually talks about these things, but the motivation for them is philosophically flawed. The idea to preserve the form of the equation

$$\mathbf{F} = m\mathbf{a}. \tag{10.81}$$

for large velocities. There are two problems with this. One is that it doesn't work, even if you allow m to depend on $\vec{v}$. You can still only preserve this form in certain special cases. But more importantly, you shouldn't want to preserve this form anyway. We have seen that the more fundamental form that arises naturally in a Lagrangian description of mechanics is

$$\mathbf{F} = \frac{d}{dt}\mathbf{p}. \tag{10.82}$$

Lorentz transformation of energy-momentum

One of the important things about energy and momentum is that they behave under Lorentz transformations very much like a space and time interval. Remember that this has to do with what happens to their values when we go from one reference frame to another. The easiest way to understand what happens to energy-momentum is to imagine that the particle has a clock on it and consider the space-time interval between two ticks of a particle's clock. A space-time interval between two events has a time component, Δt, that is the time that elapses between the two events, and a space component, $\Delta \mathbf{r}$, that is the vector from the position of one event to the position of the other. In relativistic units, of course, these two components have the same dimension. If the particle is sitting still, this interval has a time component, call it $\Delta \tau$, but its space component vanishes.

Now suppose that the particle is moving with velocity $\mathbf{v}$. Then because of time dilation, the time interval between the same two ticks of the particle's clock is

$$\Delta t = \Delta \tau \frac{1}{\sqrt{1 - v^2}} \tag{10.83}$$

But then, because

$$\mathbf{v} = \frac{\Delta \vec{r}}{\Delta t} \tag{10.84}$$

We find

$$\Delta \mathbf{r} = \Delta \frac{\mathbf{r}}{\sqrt{1 - v^2}} \tag{10.85}$$

But then we can write

$$(E, \mathbf{p}) = \frac{m}{\Delta \tau}(\Delta t, \Delta \mathbf{r}) \tag{10.86}$$

But both m and τ are invariants. Thus we conclude that $(E, \vec{p}$ must transform just like $(\Delta t, \Delta \mathbf{r}$ because the two are just proportional to one another.

4-vectors and the invariant product

I hope that by this time you are getting used to the idea of changes from one inertial frame to another, and the accompanying Lorentz transformation, as a kind of 4 dimensional analog of what we do in three dimensional space when we go from one coordinate system to another. After developing the analogy between space when we go from one coordinate system to another. After developing the analogy between 3 dimensional space and 4 dimensional space time, how can we not go on to develop the analogy between 3 dimensional vectors and 4 dimensional vectors. Indeed, our treatment of the invariant interval was the first step in developing this analogy. We saw there that the invariant interval(with c=1),

$$s^2 \equiv (t_1 - t_2)^2 - (\mathbf{r}_1 - \mathbf{r}_2) \cdot (\mathbf{r}_1 - \mathbf{r}_2) \tag{10.87}$$

is a kind of length. Like the length of a 3 dimensional vector, it has the same value in all coordinate systems, but here of course, the idea of coordinate system is enlarged to include different inertial frames, moving with different velocities.

The obvious way to go further is the find analogs in spacetime for the 3 vector, $\Delta \mathbf{r}$, and for the dot product. The analog of the 3 vector is pretty obvious. It is a 4 vector, with 4 components, the first of which is the time. So a 4 vector is a quarter of numbers,

$$A = (A_0, A_1, A_2, A_3) \tag{10.88}$$

where A_0 is referred to as the time component, and A_1, A_2 and A_3 are the space components, which are the three components of a 3 vector. We will sometimes write the 4 vector as

$$A = (A_0, \mathbf{A}) \tag{10.89}$$

recognizing that the last three components form an ordinary 3 vector.

Example: The difference between the components of two spacetime events forms a 4 vector

$$\Delta r = (\Delta r_0, \Delta r_1, \Delta r_2, \Delta r_3) = (\Delta t, \Delta x, \Delta y, \Delta z) = (\Delta t, \Delta \mathbf{r}) \tag{10.90}$$

We may sometimes refer to the four components of a 4 vector using different subscripts,

$$(A_0, A_1, A_2, A_3) \rightarrow (A_t, A_x, A_y, A_z) \tag{10.91}$$

Not any quartet of coordinates is a 4 vector. What makes a 4-vector a 4-vector is that it behaves like the coordinate interval 10.90 under a change from one inertial frame to another, that is under a Lorentz transformation. Thus if A is a 4-vector , then under a Lorentz transformation to a frame moving with speed v in the x direction, the components of A go to a new set of components A' related to the first by the Lorentz transformation

$$\Delta A_1' = \gamma(\Delta A_1 - v\Delta A_0), \tag{10.92}$$

$$\Delta A_0' = \gamma(\Delta A_0 - v\Delta A_1), \tag{10.93}$$

$$\Delta A_2' = \Delta A_2, \Delta A_3' = \Delta A_3, \tag{10.94}$$

We have now seen two things that behave this way- the space-time interval and the energy-momentum. Now for the analog of the dot product. If we have two 4-vectors, A and B,

$$A = (A_0, A_1, A_2, A_3) \tag{10.95}$$

$$B = (B_0, B_1, B_2, B_3) \tag{10.96}$$

then the combination

$$A \cdot B \equiv A_0 B_0 - \mathbf{A} \cdot \mathbf{B} \tag{10.97}$$

has the same value in any frame of reference. The notation here is a bit condensed. If you see a dot product between things that don't have vector indices, you should assume that the things are 4-vectors and that the dot product is the 4-dimensional invariant product defined by 10.97. It is also obviously unchanged by rotations because the space vectors enter only through the ordinary dot product, which is unchanged by rotations. Note also that if we set B=A, we recover the equation for the invariant interval, s^2, and therefore we can write the invariant interval as

$$s^2 = \Delta r \cdot \Delta r \tag{10.98}$$

just as for 3 dimensional vectors, the square of the distance between two vectors is a dot product,

$$l^2 = \Delta \mathbf{r} \cdot \Delta \mathbf{r} \tag{10.99}$$

Right now, 4-vectors and the invariant product probably look like just a pretty mathematical analogy. But we will see shortly when we talk about energy and momentum that they are an indispensible part of our too box for dealing with the relativistic world.

Energy, Momentum, Velocity and Mass

There are several ways of writing the relation between energy, momentum, velocity and mass. The one that we started with,

$$E = \frac{m}{\sqrt{1 - v^2}}, \mathbf{p} = \frac{m\mathbf{v}}{\sqrt{1 - v^2}}. \tag{10.100}$$

is actually not the most useful, because it doesn't make sense for massless parti-cles, such as the particles of light itself, because both the numerator and the denom-inators vanish as $m \to 0$. However, we can combine these into two relations that are even more general, and make sense for any m. First consider the obvious one-form the invariant product of the energy-momentum with itself. Explicit calculation gives

$$E^2 - \mathbf{p}^2 = m^2. \tag{10.101}$$

As expected, the dependence on v has gone away, because the invariant on the left hand side does not depend on the inertial frame, and thus cannot depend on how fast the particle is moving.

We can also get a relation that makes sense as $m \to 0$ by dividing the expression for momentum by the expression for energy.

$$\mathbf{v} = \frac{\mathbf{p}}{E}. \tag{10.102}$$

These two relations are the most general formulation of the relations among energy, momentum, velocity and mass.

Massless particles

When $m = 0$, (40) and (41) are perfectly sensible, but the result is a bit odd. For $m = 0$, implies that

$$|\mathbf{p}| = E. \tag{10.103}$$

Then (41) implies that

$$\mathbf{v} = \frac{\mathbf{p}}{|\mathbf{p}|} = \hat{p}. \qquad (10.104)$$

which means that a massless particle always travel at the speed of light. If you think about this for a minute, it really makes your head hurt. For one thing, it means that there is no way that the state of such a particle can be specified only by its position, because massless particles with different energies move at the same speed, and thus cannot be distinguished by their speed. Thus unlike the classical particles you are used to be, massless particle with different energy can have exactly the same trajectories. The resolution of this puzzle classically is to say that one shouldn't talk about massless particle at all, but just about classical waves like the electromagnetic waves. But quantum mechanically, these particles really exist.

General addition of velocities

A number of people in their early evaluations asked for more involved examples. So before we get to particle collisions, we will discuss how to use the invariant product to do addition of velocities in three dimensions. That is the general addition of velocity. Suppose that we see in our frame two objects with velocities $\mathbf{v}_1$ and $\mathbf{v}_2$. In the frame of reference of one of these objects, what is the speed of the other? We sometimes express this question somewhat less precisely by asking how fast does one of the objects see the other moving, without exactly defining what we mean by see. In fact, we don't really mean see in the conventional sense. Rather we imagine that there are observers at rest with respect to object 1 with syncronized clocks, and they make measurements of the times at which object 2 passes their positions. Then they get together and calculate object 2's speed. This is not seeing, but at least we know exactly what we mean.

We could solve this problem the same way we did addition of velocities in one dimension using the Lorentz transformation. But instead, let's use the invariant product. Suppose object 1 has mass m_1 and particle 2 has mass m_2. Then the 4 momenta of the two objects are

$$\mathbf{P}_1 = (m_1\gamma_1, m_1\mathbf{v}_1\gamma_1), \mathbf{P}_2 = (m_2\gamma_2, m_2\mathbf{v}_2\gamma_2) \tag{10.105}$$

where

$$\gamma_1 = \frac{1}{\sqrt{1 - v_1^2}}, \gamma_2 = \frac{1}{\sqrt{1 - v_2^2}} \tag{10.106}$$

and the invariant product is

$$P_1 \cdot P_2 = (m_1\gamma_1)(m_2\gamma_2) - (m_1\mathbf{v}_1\gamma_1) \cdot (m_2\mathbf{v}_2\gamma_2) = m_1 m_2 \gamma_1 \gamma_2 (1 - \mathbf{v}_1 \mathbf{v}_2) \tag{10.107}$$

In the frame of reference of object 1, the momenta are

$$P_1' = (m_1, 0) and P_2' = (m_2\gamma, m_2\mathbf{v}\gamma) \tag{10.108}$$

where $\mathbf{v}$ and the velocity and γ factor of object 2 in the frame of reference of object 1, which is what we want to know. We can compute the invariant product again

$$P_1 \cdot P_2 = P_1' \cdot P_2' = (m_1)(m_2\gamma) - (0) \cdot (m_2\mathbf{v}\gamma) = m_1 m_2 \gamma \tag{10.109}$$

Then we have

$$m_1 m_2 \gamma = m_1 m_2 \gamma_1 \gamma_2 (1 - \mathbf{v}_1 \cdot \mathbf{v}_2) \tag{10.110}$$

The masses play no role

$$\gamma = \gamma_1 \gamma_2 (1 - \mathbf{v}_1 \cdot \mathbf{v}_2) \tag{10.111}$$

Now we have done all the physics. The rest is just algebra. Squaring and substituting for the γ gives

$$\frac{1}{1 - v^2} = (\frac{1}{1 - v_1^2})(\frac{1}{1 - v_2^2})(1 - \mathbf{v}_1 \cdot \mathbf{v}_2)^2 \tag{10.112}$$

Inverting gives

$$1 - v^2 = \frac{(1 - v_1^2)(1 - v_2^2)}{(1 - \mathbf{v}_1 \cdot \mathbf{v}_2)^2} \tag{10.113}$$

$$v^2 = 1 - \frac{(1 - v_1^2)(1 - v_2^2)}{(1 - \mathbf{v}_1 \cdot \mathbf{v}_2)^2} \tag{10.114}$$

$$= \frac{(\mathbf{v}_1 - \mathbf{v}_2) \cdot (\mathbf{v}_1 - \mathbf{v}_2) + (\mathbf{v}_1 \cdot \mathbf{v}_2)^2 - v_1^2 v_2^2}{(1 - \mathbf{v}_1 \cdot \mathbf{v}_2)^2} \tag{10.115}$$

This looks complicated, but notice that if we put the factors of c back in it is

$$= \frac{(\mathbf{v}_1 - \mathbf{v}_2) \cdot (\mathbf{v}_1 - \mathbf{v}_2) + (\mathbf{v}_1 \cdot \mathbf{v}_2)^2/c^2 - v_1^2 v_2^2/c^2}{(1 - \mathbf{v}_1 \cdot \mathbf{v}_2)^2} \tag{10.116}$$

You see that for small velocities it goes to what we expect in the nonrelativistic limit. You can also check that if $\mathbf{v}_1$ and $\mathbf{v}_2$ are parallel, this reduces to the result we got last week. Note further that we have only discussed the square of the velocity, which it the square of the speed -ignoring the vector nature of $\mathbf{v}$. There is a very good reason for this. In the general case, there is no particularly natural coordinate system in which to define the direction of this vector. All we can really talk about in an invariant way is the magnitude of v.

Note also that 10.114 gives a particularly simple way of seeing that the speed of light is the same in all reference frames. If either v_1 and v_2 is 1, then the second term on the right hand side vanishes, and $v = 1$ as well. This means that a particle moving with the speed of light in the lab frame is also moving at the speed of light in the frame of reference of any massive particle.

You can think about the crucial relation if you like as a kind of interpretation of the invariant product, analogous to the statement that the cosine of the angle between two three dimensional vectors $\mathbf{r}_1$ and $\mathbf{r}_2$ can be computed as

$$\cos\theta = \frac{\mathbf{r}_1 \cdot \mathbf{r}_2}{\sqrt{(\mathbf{r}_1 \cdot \mathbf{r}_1)(\mathbf{r}_2 \cdot \mathbf{r}_2)}} \tag{10.117}$$

If we use the fact that

$$P_1 \cdot P_1 = m_1^2, P_2 \cdot P_2 = m_2^2 \tag{10.118}$$

we can rewrite 10.109 as a corresponding relation for time-like 4-vectors P_1 and P_2, both with positive energy, and the γ factor that depends on the velocity of object 1 in the frame of object 2.

$$\gamma = \frac{P_1 \cdot P_2}{(P_1 \cdot P_2)(P_1 \cdot P_2)} \tag{10.119}$$

This looks even more analogous to 10.117 if we write γ in terms of the rapidity.

$$\tanh\phi = \beta = \frac{v}{c} \tag{10.120}$$

and

$$\gamma \equiv \frac{1}{\sqrt{1-\beta^2}} = \frac{1}{\sqrt{1-\tanh^2\phi}} = \cosh\phi \tag{10.121}$$

so that

$$\cosh\phi = \frac{P_1 \cdot P_2}{(P_1 \cdot P_1)(P_2 \cdot P_2)} \tag{10.122}$$

Instead of the cosine of the angle that we get with a dot product, with the invariant product, we get the hyperbolic cosine of the rapidity.

10.7   Particle collisions

The relativistic energy and momentum are incredibly important. We derived these expressions by thinking about single free particles, but they are much more generally useful. The energy and momentum are conserved in all collisions of small particles, even when new particles are created or when particles initially present are annihilated.

Conservation means simply that when we add up the energies and momenta of the particles in the initial state of some scattering process the result is the same as if we add up the energies and momenta of the particles in the final state. The thing that I want to try to convince you of today is that it is much easier to determine the constraints that come from energy and momentum conservation if we think of the energy and momentum as a 4-vector and use the fact that the dot product of 4-vector is independent of the frame.

Here we discuss some examples of the use of conservation of the energy-momentum 4-vector to analyze the decay, scattering, and production of particles. There is a very simple general idea that underlies all of these problems. The idea is to get rid of things that you don't know by using the relation $E^2 - \vec{p}^2 = m^2$.

K^+ decay

There is a particle called K^+. It is called a strange particle for historical reasons. This is not because it is peculiar, but because it carries a property called strangeness. It decays rather quickly into a pair of pions. Pions are the lightest of the particles made out of quarks and antiquarks so they show up often. The K^+ can decay into one neutral pion (π^0), which has a mass of about $m_{\pi+} \approx 135\text{MeV}$ and one charged pion (π^+), which has a mass of about $m_{\pi+} \approx 140\text{MeV}$. The K^+ has a mass of $m_{K+} \approx 494\text{MeV}$. Now suppose that the dacay of the K^+ occurs at rest. What are the energies of the two pions? This is typical sort of question in what might be called decay kinematics. To answer such questions, we think about 4-vectors and use conservation of energy and momentum. Let us call the energy-momentum 4-vectors K for the K^+ and π^+ and π^0 for the π^+ and π^0 respectively. Conservation of energy and momentum is the statement that the 4-dimensional vectors satisfy

$$K = \pi^+ + \pi^0 \tag{10.123}$$

Note that this is true in any frame of reference. In the rest frame, the 4-vectors look like

$$K = (m_K, 0) \tag{10.124}$$

$$\pi^+ = (E_+, \mathbf{p}_+) \tag{10.125}$$

$$\pi^0 = (E_0, \mathbf{p}_0) \tag{10.126}$$

We have used a standard trick here - one that you should be familiar with from our rules of coherence. If you don't know something, give it a name. Now 10.123 can be used to say things about 10.125, for example, $\mathbf{p}_+ = -\mathbf{p}_0$. In problems like this, it is often convenient to manipulate the 4-vectors sympolically for a while before actually doing the calculation. The idea of such manupulations is to be able to use the scalar product to compute what you want to know without having to calculate what we don't care about. Here for example, suppose that we first want to calculate the energy of the π^+, which we have called E_+. If we could calculate the value of the scalar product $K \cdot \pi_+ = m_K E_+$, that would immedately give us E_+. so suppose that we rewrite 10.123 as

$$K - \pi_+ = \pi_0 \tag{10.127}$$

Now if we take the scalar product of each side of this equality with itself, we will get terms involving $K \cdot \pi_+ = m_K E_+$:

$$(K - \pi_+) \cdot (K - \pi_+) = K \cdot K - 2K \cdot \pi_+ + \pi_+ \cdot \pi_+ \tag{10.128}$$

$$= m_K^2 - 2K \cdot \pi_+ + m_{\pi^+}^2 = \pi_0 \cdot \pi_0 = m_{\pi^0}^2 \qquad (10.129)$$

Now we can solve this for $K \cdot \pi_+$

$$K \cdot \pi_+ = \frac{m_K^2 + m_{\pi^+}^2 - m_{\pi^0}^2}{2} \qquad (10.130)$$

Or

$$E_+ = \frac{m_K^2 + m_{\pi^+}^2 - m_{\pi^0}^2}{2m_K} \qquad (10.131)$$

Now we can get E_0 either by repeating the same calculation with + and 0 interchanged, or by using energy conservation. The result is

$$E_0 = \frac{m_K^2 + m_{\pi^0}^2 - m_{\pi^+}^2}{2m_K} \qquad (10.132)$$

If you were asked to do so, you could now go on and calculate the magnitudes of the momenta of the pions by using $E^2 - \vec{p}^2 = m^2$. From there, you could calculate the speeds, although this is seldom very interesting in such collisions. You cannot calculate the direction of the momentum or velocity, because this is actually quite random. The K^+ is a particle with no angular momentum. When it is at rest, there is no vector associated with it. And therefore there is no direction picked out for its decay products. They go off at random with equal probability in all directions.

Neutrino scattering

Neutrinos are very light particles. Until recently, we thought that they might be massless, like photons. But it now appears that the neutrinos have tiny masses.

Furthermore, these masses are very peculiar. There are three different kinds of neutrinos: $\nu_e, \nu_\mu, and \nu_\tau$. The names refer to the processes in which these neutrinos are produced, which involve respectively the electron and heavier versions of the electron, the μ and the τ. The tiny masses do not respect these distinctions, and they produce bizarre quantum mechanical mixing between these different types of neutrinos. But if it were not for these weird quantum mechanical effects, we could ignore the neutrino masses altogether. We will simply pretend that neutrinos are massless, which is an excellent approximation for the sort of questions that we can ask and answer.

In spite of the fact that neutrinos have no electric charge and almost no mass, it is possible to make beams of neutrinos. With these beams, we can observe processes such as the scattering of a ν_μ with energy E from an electron at rest to produce a final state consisting of a ν_e and a μ. The μ has a mass m_μ about 207 times the mass of the electron, m_e.

Now a question that you might ask about this process is the following. Suppose that you see a ν_e in the final state flying off at an angle ϕ from the initial direction of the ν_μ: What does energy and momentum conservation tell you about the energy and angle of the ν_e in the final state?

The 4-vectors look like

$$e = (m_e, 0, 0, 0) \tag{10.133}$$

$$\nu_\mu = (E, E, 0, 0) \tag{10.134}$$

$$\nu_e = (E_1, E_1 \cos \phi, E_1 \sin \phi, 0) \tag{10.135}$$

$$\mu = (E_2, p_2 \cos \theta, -p_2 \sin \theta, 0) \tag{10.136}$$

Some comments about this is in order. I have chosen to put the initial ν_μ momentum along the x axis. That is no problem, because I can rotate my coordinate system to make it so. Likewise, I have assumed that the scattering takes place in the $x - y$ plane, which I can again do by just rotating the lengths of their momentum vectors to equal their energies. If I have not done this and just given the momenta names, we would have quickly gotten to this point when we imposed $E^2 - \mathbf{p}^2 = m^2 = 0$ on these 4-momenta. I have also imposed a little bit of energy momentum conservation by writing μ in the $x - y$ plane. Again if we had put in a z component for μ, we would have quickly realized that it must be zero because all the other 4-vectors have zero z component by construction.

Now we could simply impose energy and momentum conservation of the rest of 10.135, and try to solve the equations. But it is better to think. For example, we can easily find E_1, because energy-momentum conservation

$$\nu_\mu + e = \mu + \nu_e \tag{10.137}$$

implies

$$\nu_\mu + e - \mu_e = \mu \tag{10.138}$$

This is a good way to write things, because when we take the scalar product of each side with itself, all the nonsense in μ, drops out, and we can write

$$(\nu_\mu + e - \nu_e) \cdot (\nu_\mu + e - \nu_e) = \mu \cdot \mu = m_\mu^2 \qquad (10.139)$$

The left hand side of 10.139 is

$$\nu_\mu \cdot \nu_\mu + e \cdot e + \nu_e \cdot \nu_e + 2\nu_\mu \cdot e - 2\nu_\mu\nu_e - 2e \cdot \nu_e. \qquad (10.140)$$

which with 10.139 implies

$$m_e^2 + 2m_e E - 2EE_1(1 - \cos\phi) - 2m_e E_1 = m_\mu^2 \qquad (10.141)$$

so that

$$E_1 = \frac{m_e^2 + 2m_e E - m_\mu^2}{2E(1 - \cos\phi) - 2m_e} \qquad (10.142)$$

Practically speaking, this is a bit of a swindle. There is nothing wrong with the calculation above, but a particle physicist would never ask you to calculate things in terms of the angle of the final state neutrino. This is because neutrinos are very hard to see. They very seldom interact with anything. But there are many ways of making the direction of charged particle track show up, all making use of the electric charge and its interactions. So it would make more sense physically to ask you to find things in terms of the μ angle θ, or its energy, E_2, or momentum, p_2. This is a little more involved algebraically, but the particle is the same.

There is an interesting and useful class of questions in which the kinematics does not completely fix the interesting quantities, and you have to think about how to make them bigger or smaller. Here is a simple example. Suppose that you hit a proton at rest with a proton of energy E. What is the minimum energy required to produce a final state consisting of three protons, each with mass m_p, and one antiproton? More generally, one might ask what minimum energy is required to produce any particular final state?

Let us begin by discussing the question in general. The key to problems like this is to treat the whole final state as a single entity. The final state, whatever it is , will have some total energy-momentum 4-vector

$$T = (E_T, \mathbf{P}_T) \tag{10.143}$$

Suppose that

$$E_T^2 - \mathbf{P}_T^2 = M_T^2 \tag{10.144}$$

Then in terms of M_T, this problem is formally equivalent to problem of producing a particle with mass M_T by colliding a proton of energy E with a proton at rest. Define the proton 4-momenta as

$$P_1 = (E, p), P_2 = (m_p, 0) \tag{10.145}$$

Then 4-momentum conservation implies

$$P_1 + P_2 = T \tag{10.146}$$

Taking the invariant product of each side gives

$$2m_p^2 + 2m_pE = M_T^2 \tag{10.147}$$

$$E = \frac{M_T^2 - 2m_p^2}{2m_p} \tag{10.148}$$

Evidently, to minimize E we need to minimize M_T. So how do we do that?

Here is a mathematical issue that is clearly related. Given a set of 4-vectors describing some number of particles, $(E_j, \mathbf{P}_j)$, with $E_j \geq |\mathbf{P}_j|$ such that

$$E_T = \Sigma_j E_j, \mathbf{P} = \Sigma_j \mathbf{P}_j \tag{10.149}$$

, where $m_j^2 \equiv E_j^2 - \mathbf{P}_j^2 \geq 0$, what is the minimum value of the total mass M_T defined 10.144?

Let us do this by going to the frame of reference in which $\mathbf{P}'_T = 0$, which we can always do -this frame moves with velocity $\frac{\mathbf{P}_T}{E_t}$ with respect to the original frame. In this frame, M_T is just the sum of the energies, E_j, which in turn are bounded by the masses, m_j. Thus the smallest we can possibly hope to make M_T is $\Sigma_j m_j$. Now let us show that we can always actually do this, or at least come arbitrarily close to it. If there are no massless particles in the final state, it is easy, because we can take all the massive particle to have zero momentum. Then $E_j = m_j$, and $M_T = \Sigma_j m_j$, If there are massless particles, we cannot take their momenta to be zero, but we can take them to be very small. In this way we can come arbitrarily close to the theoretical minimum. If we now boost this final state by a Lorentz transformation, all of the massive particles will be travelling with the same velocity, and the massless particles will still have arbitrarily small energy and momentum.

Thus in our question about the final state of three protons and one antiproton, because all of the final state particles are massive, we can take all four particles to be moving with the same velocity, v, and the total mass, $M_T = 4m_p$. Putting this into 10.148 gives the result:

$$E = \frac{M_T^2 - 2m_p^2}{2m_p} = \frac{16m_p^2 - 2m_p^2}{2m_p} = 7m_p \tag{10.150}$$

The idea is to think of whole collection of particles as having a mass- computed from the total energy momentum 4-vector. This is analogous to a very simple problem in geometry. Suppose you have a set of vectors whose lengths you know but whose directions are variable. How do you choose the directions of your vectors so that the sum of the vectors has the maximum possible length? The answer is that you want to take all the vectors in the same direction. The velocity of the particle is analogous to the direction of the vector. But what you see once again is that the minus sign in the variant interval makes things more complicated.

μ decay

We have talked about the fact that the heavy version of the electron called the muon, μ, is unstable. It decays into an electron, a muon neutrino, and an electron antineutrino:

$$\mu^- \to e^- + \nu_\mu \bar{\nu}_e \tag{10.151}$$

This process is actually very closely related to the inverse of the scattering process we discussed earlier.

$$\nu_e + \mu^- \rightarrow \nu_\mu + e^- \qquad (10.152)$$

There is a sense in which the $\bar{\nu}_e$ in 10.151 is related to a ν_e traveling backward in time. That is to say that a $\bar{\nu}_e$ in the final state is related to a ν_e in the initial state. If we made this change in 10.151, we would get just the inverse of the process 10.152. In some sense, this is why antiparticles must exist for every type of particle. This is actually related to a discrete symmetry called CPT which stands for **Charge Conjugation-Parity-Time reversal** which at least so far seems to be an exact symmetry of the world.

Here we can ask some questions about this. Suppose that the μ decays while it is at rest.

1. What is the minimum possible energy of the electron?

The smallest energy the electron could possibly have is m_e, which it would have if it were at rest. Is this possible? Yes, We can conserve energy and momentum if all the rest of the energy goes into two back-to-back neutrinos, so the 4-momenta would look like

$$\mu = (m_\mu, 0), e = (m_e, 0), \nu_\mu = (E, E\hat{v}), \bar{\nu}_e = (E, -E\hat{v}). \qquad (10.153)$$

Conservation of energy and momentum works if $E = (m_\mu - m_e)/2$.

2. What is the maximum possible energy of the electron?

To get the maximum possible energy for the electron, what we want is that the effective mass of the rest of the stuff in the decay, the two neutrinos, should be as small as possible. But if the two neutrinos have parallel momenta, the mass of the two-neutrino system is zero, That is the best we can do. The process then looks like

$$\mu = (m_\mu, 0), e = (E, p\hat{v}), \nu_\mu = (xp, -xp\hat{v}), \bar{\nu}_e = (yp, -yp\hat{v}). \tag{10.154}$$

where $x + y = 1$. We can calculate E easily as we have done in other problems if we note that $\mu - e$ has mass 0, so that

$$m_\mu^2 - 2m_\mu E + m_e^2 = 0 \rightarrow E = \frac{m_\mu^2 + m_e^2}{2m_\mu} \tag{10.155}$$

3. What is the minimum possible energy of one of the neutrino?

Either of the neutrinos can have arbitrarily small energy and momentum. There is enough freedom to satisfy energy and momentum conservation with the other two carrying the load.

4. What is the maximum possible energy of one of the neutrion?

This is actually related to the previous question. The maximum neutrino energy arises when the neutrino recoils against the minimum possible mass, which is the electron mass, with the other neutrino carrying negligible energy and momentum:

$$\mu = (m_\mu, 0), e = (E, p\hat{v}), \nu_\mu = (p, -p\hat{v}), \bar{\nu}_e = (\approx 0, \approx 0). \tag{10.156}$$

Now the mass of the 4-vector $\mu - \nu_\mu$ is μ_e, so

$$m_\mu^2 - 2m_\mu p = m_e^2 \rightarrow p = \frac{m_\mu^2 - m_e^2}{2m_\mu} \tag{10.157}$$

5. What is the miximum mass of the two neutrino system?

We have really done this one already. The two-neutrino system has its maximum mass when it and the electron are both at rest, and the mass of the two-neutrino

system is $m_\mu - m_e$.

6. What is the minimum mass of the two neutrino system?

We've done this one also. If the two neutrino momenta are parallel, the mass is

zero.

Colliders

Here is an interesting and important process

$$e^- + e^+ \rightarrow J/\psi \tag{10.158}$$

where e^- is an electron, e^+ is a positron, the antiparticle of the electron, and J/ψ

is a particle that has two names for historical reasons. It was discovered allegedly

independently at Brookhaven and at SLAC. Anyway, it has a mass of $m_{J/\psi} \approx 3097$

MeV. Now suppose that we let positrons with energy E collide with electrons at rest

to produce J/ψs. What energy is required? The 4-vectors are as follows:

$$e^- = (m_e, 0) e^+ = (E, \mathbf{p}) J/\psi = (E', \mathbf{p'}) \tag{10.159}$$

Now we can use energy-momentum conservation

$$e^- + e^+ = J/\psi \tag{10.160}$$

Taking the scalar product of each side with itself gives

$$m_e^2 + 2m_e E + m_e^2 = m_{J/\psi}^2 \tag{10.161}$$

so that

$$E = \frac{m_{J/\psi}^2 - 2m_e^2}{2m_e} \approx 9385 GeV \tag{10.162}$$

This is a huge energy scale, beyond what is presently available at accelarators. The problem is that the electron is very light. A collision between a very high energy electron and an electron at rest is a bit like a nonrelativistic collision between a moving truck and a feather. Very little energy is actually transfered in such a collision.

However, it is much easier to produce the J/ψ in a collider, in which an electron and positron collide with equal and opposite velocities and momenta. In this case, the 4-vectors look like

$$e^- = (E, \mathbf{p}) e^+ = (E, -\vec{p}) J/\psi = (m_{J/\psi}, 0) \tag{10.163}$$

Here energy-momentum conservation implies

$$E = m_{J/\psi}/2 \approx 1.55 GeV \tag{10.164}$$

Colliding beams, in this case, make a huge difference. The difference is that in the collision with a particle at rest, much of the energy of the incoming particle is wasted producing kinetic energy of the collision products. This effect exists in Newtonian physics also, but it is much worse at high energies where relativistic physics takes over. Note that the problem is particularly bad for the electron, because it is the lightest particle with electric charge. As 10.162 shows, the energy required to produce a heavy particle of mass M in a fixed target collision between particles of mass m, much less than M is inversely proportional to m, so the lightness of the electron is a terrible problem. But if you want to have the capability to produce the heaviest possible things, colliding beams are essentially no matter what you are colliding.

Why is this interesting? What are these heavy particles that particle physicists make, and why would you want to make them? The first thing to say is that we don't need heavy particles to make heavy things. The things in the universe that are much heavier than protons and neutrons and electrons are made by putting many copies of these light things together. This can be done in a practically infinite number of way and produces lots of interesting physics. But the heavy particles of particle physics are very different. They are not just light things put together in interesting ways. And there are only a very few of them, with completely well difined masses and properties. To make one of these particles with mass M, we must not only have enough energy, M, in the zero momentum frame, but we must also concentrate that energy in a tiny region of space and time, of size $1/M$. They are genuinely new indivisable degrees of freedom that appear only when look at the world at large energies and small distances. They are not useful, unless you want to know how the world works.

Classical Mechanics

저자 정태성(Taeseong Jeong)

초판 발행 2021년 9월 1일

지은이   정태성
펴낸이   정주택
펴낸곳   도서출판 코스모스
등록번호 414-94-09586
주소 충북 청주시 서원구 신율로 13
전화 043-234-7027
팩스 050-7535-7027

ISBN 979-11-91926-00-2

값 20,000원